THE RIBONUCLEOTIDE REDUCTASE FAMILY: GENETICS AND GENOMICS

THE RIBONUCLEOTIDE REDUCTASE FAMILY: GENETICS AND GENOMICS

EDUARD TORRENTS
MARGARETA SAHLIN
AND
BRITT-MARIE SJÖBERG

Nova Biomedical Books
New York

Copyright © 2009 by Nova Science Publishers, Inc.

All rights reserved. No part of this book may be reproduced, stored in a retrieval system or transmitted in any form or by any means: electronic, electrostatic, magnetic, tape, mechanical photocopying, recording or otherwise without the written permission of the Publisher.

For permission to use material from this book please contact us:
Telephone 631-231-7269; Fax 631-231-8175
Web Site: http://www.novapublishers.com

NOTICE TO THE READER

The Publisher has taken reasonable care in the preparation of this book, but makes no expressed or implied warranty of any kind and assumes no responsibility for any errors or omissions. No liability is assumed for incidental or consequential damages in connection with or arising out of information contained in this book. The Publisher shall not be liable for any special, consequential, or exemplary damages resulting, in whole or in part, from the readers' use of, or reliance upon, this material.

Independent verification should be sought for any data, advice or recommendations contained in this book. In addition, no responsibility is assumed by the publisher for any injury and/or damage to persons or property arising from any methods, products, instructions, ideas or otherwise contained in this publication.

This publication is designed to provide accurate and authoritative information with regard to the subject matter covered herein. It is sold with the clear understanding that the Publisher is not engaged in rendering legal or any other professional services. If legal or any other expert assistance is required, the services of a competent person should be sought. FROM A DECLARATION OF PARTICIPANTS JOINTLY ADOPTED BY A COMMITTEE OF THE AMERICAN BAR ASSOCIATION AND A COMMITTEE OF PUBLISHERS.

Library of Congress Cataloging-in-Publication Data

ISBN: 978-1-60692-419-8

Available upon request

Published by Nova Science Publishers, Inc. ✢ New York

Contents

Preface		vii
Chapter I	Introduction	1
Chapter II	RNR Classification and Occurrence	3
Chapter III	RNR Diversification	19
Chapter IV	Regulation of RNRs	23
Chapter V	Mutational Studies in RNRs	39
Chapter VI	RNR as an Antiproliferative Target for Disease Control	47
Chapter VII	Future Antiproliferative Regimes	61
Acknowledgments		63
References		65
Index		89

Preface

Ribonucleotide reductase (RNR), a universal enzyme present in essentially all living cells and organisms, has a central role in DNA replication and repair by catalyzing production of deoxyribonucleotides from the corresponding ribonucleotides. Three major classes of RNRs are known, differing in their cofactor requirements: class I RNRs (with subclasses Ia and Ib) carry a stable tyrosyl radical and are oxygen-dependent, class II RNRs require the vitamin B_{12} cofactor 5'-deoxyadenosylcobalamin and are oxygen-independent, and class III RNRs carry a stable glycyl radical and are oxygen-sensitive. Despite these differences, all classes have a similar reaction mechanism and the same highly specific catalytic core structure, indicating that they evolved from a common ancestor. Biochemical studies of RNRs from selected model organisms in combination with the vast number of deduced RNR sequences from publicly available complete genomic sequences show that whereas eukaryotes and their viruses with few exceptions contain only class Ia RNRs, all three major RNR classes are found among prokaryotes and bacteriophages and quite often one organism encodes more than one class of RNR. They are compiled in an open access database, called RNRdb for Ribonucleotide Reductase database that is available at http://rnrdb.molbio.su.se. RNRs are produced in a strictly controlled way depending upon growth phase and environmental cues. We describe a comprehensive summary of how the expression of RNR genes is regulated in several eubacterial organisms and in yeast. Due to RNR's importance for the realization of DNA replication, it has been recognized as a possible target for antiproliferative therapy. We present a comprehensive summary of RNR-specific inhibitors that have reached clinical trials and/or are currently used in clinical therapy.

Chapter I

Introduction

Ribonucleotide reductase (RNR) is a universal enzyme present in essentially all living cells and organisms. It catalyzes the production of building blocks for DNA by reducing ribonucleotides to deoxyribonucleotides. The first RNR enzymes were isolated and characterized in the 1960ies [1-5], and the first RNR operon to be cloned and sequenced was reported in 1984 [6]. This nucleotide sequence was obtained from a recombinant plasmid that could complement a mutant *Escherichia coli* with a conditional lethal RNR defect. Later RNR genes were also cloned from mouse [7, 8] and yeast [9, 10] by similar methodology, or by using known peptide sequences from highly purified RNRs to synthesize degenerate oligonucleotide primers for cloning and sequencing. A burst of genomic DNA sequences that code for RNRs have appeared since 1995 and onwards, as the methods for whole genome sequencing were worked out and further developed. By now (October 2007) the public databases list completely sequenced genomes for 45 Archaea, more than 500 Eubacteria, and approximately 35 Eukaryota, mainly fungi but also plants and animals including several mammals and man. In addition, the public databases also include massive numbers of sequenced genomes from bacterial and eukaryotic viruses, some of which carry their own genes for RNRs. Regulation of RNR genes were studied initially by observing alterations in the expression pattern of individual genes in different mutant backgrounds affecting potential regulators and transcription factors, and by mapping binding sites for individual regulating factors to the vicinity of the RNR genes. Recent high-throughput methods to study global gene expression patterns and proteomic studies have offered novel insight into the regulation of RNR genes, and such studies will undoubtedly revolutionize our understanding of how organisms control their production of deoxyribonucleotides for DNA replication and repair. The central role of RNR in production of building blocks for all forms of DNA synthesis makes it a potential candidate for antiproliferative therapy. The first antiproliferative drug shown to be a specific inhibitor of RNR enzyme activity came in clinical use in the late 1960ies [11-13]. Similar avenues based on knowledge of RNR function have been explored extensively and a number of therapies directed towards RNR have found their way to clinical trials and in some cases clinical use. This chapter deals with the current classification of RNRs, which types of RNRs that are encoded by different groups of organisms, how RNR genes are regulated in a few studied microorganisms and a summary of drugs and therapies used to inactivate the function of RNR. For information on specific aspects of RNRs, we refer to the subsequent chapters in this volume and to some recent reviews [14-22].

Chapter II

RNR Classification and Occurrence

Currently three different classes of RNR are known. The classification of RNRs is based on cofactor requirements. Class I RNRs require a diiron-oxo site usually coupled to a stable tyrosyl radical, class II RNRs require a vitamin B_{12} coenzyme (5'-deoxyadenosylcobalamin), and class III RNRs require a stable glycyl radical. Each cofactor mediates formation of a transient thiyl radical at a conserved cysteine residue in the active site (Figure 1, upper part), the diiron-oxo/tyrosyl radical site and the glycyl radical site by radical transfer, and the cobalamin by homolytic cleavage followed by radical transfer. The so formed thiyl radical at the active site promotes 3' hydrogen abstraction from the bound substrate (Figure 1, lower part). After this first one-electron oxidation the 2'-OH group leaves as water and the substrate undergoes the two-electron reduction proper, and thereafter a one-electron reduction by reintroduction of the abstracted 3' hydrogen atom. This step regenerates the transient thiyl radical, which in turn regenerates the diiron-oxo/tyrosyl radical site in class I RNRs, the cobalamin in class II RNRs, or the glycyl radical in class III RNRs. Apart from the differences in cofactor requirements each RNR class has several other characteristics. Interestingly, and at the time unexpectedly, all three RNR classes have a common core structure comprising the active site region [23-25]. It consists of a 10-stranded β/α-barrel that is wide enough to allow protrusion of an extended loop with the cysteine that transiently harbors the thiyl radical at its tip (see Chapters 3 and 9). Domains that are specific for each RNR class are added to this core structure, and we will come back to this point below.

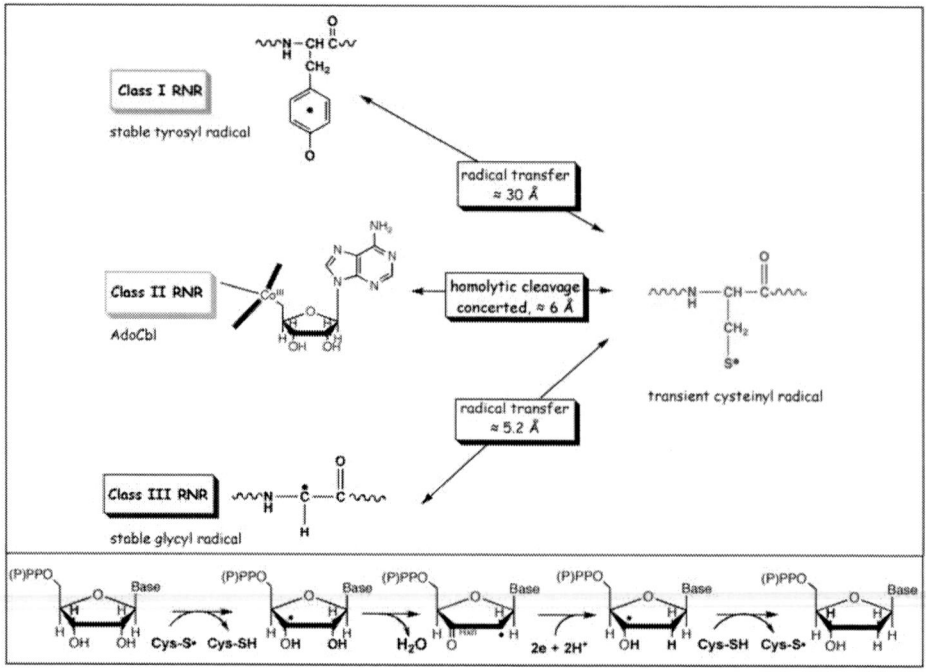

Figure 1. Generation of the transient active site radical in RNRs and a schematic reaction mechanism. The three different classes of RNRs utilize different cofactor functions to generate the essential transient thiyl radical at the active site (upper part), and follow a common reaction mechanism (lower part).

Class I RNRs

Class I RNRs are built of two non-identical polypeptides, α and β. In general, the quaternary structure of class I RNRs is a dimer-of-dimers ($\alpha_2\beta_2$), but several different oligomeric forms have been described lately [26-28]. The longer α_2 protein (also denoted NrdA or R1, see table 1 for a summary of the RNR nomenclature) comprises the active site region with three conserved cysteine residues [29, 30], whereas the shorter β_2 protein (also denoted NrdB or R2) harbors the diiron-oxo site and the tyrosyl radical [31, 32]. The active site and the diiron-oxo/tyrosyl-radical site are more than 30Å apart in the three-dimensional structure of the class I RNR [23, 33].

The NrdA protein also harbors two different types of allosteric sites. One of the allosteric sites (the specificity site) regulates the substrate specificity, i.e. which of the four possible substrates will be preferentially reduced, and the other allosteric site (the overall activity site) regulates the overall activity of the enzyme; i.e. whether it is activated or inhibited [14]. ATP, dTTP, dGTP and dATP regulate the substrate specificity, and ATP (activator) and dATP (inhibitor) regulate the overall activity. The physiological substrates for class I RNRs are the ribonucleoside diphosphates CDP, UDP, GDP and ADP, and in general ATP or low concentrations of dATP (binding only to the specificity site) activate CDP and UDP reduction, dTTP activates GDP reduction and dGTP activates ADP reduction. ATP bound to the overall activity site is a general positive effector for catalysis, whereas high concentrations of dATP binding also to the overall activity site is a general negative effector for catalysis. The specificity site is formed by flexible loops interacting at the dimer interface of the NrdA protein, and the overall activity site is in a separate N-terminal domain [23] called the ATP-cone, a structural motif consisting of three β–strands and four α–helices and found in many different proteins that bind ATP and dATP [34]. These allosteric principles and domains are the same for all RNR classes, although the overall activity regulation/ATP-cone is missing from some (see below).

Class I RNRs are oxygen requiring because generation of the essential tyrosyl radical requires oxygen-dependent oxidation of the diiron site [31]. Once the tyrosyl radical is formed, class I RNRs can perform multiple turnovers in absence of oxygen depending upon the stability of the tyrosyl radical and the diiron-oxo site. However, class I RNRs are only found in organisms that can grow aerobically, e.g. almost all eukaryotes and quite a number of prokaryotes, primarily Eubacteria.

Eubacterial class I RNRs are further subdivided into class Ia and class Ib [14]. In class Ia the *nrdA* gene encodes the α–polypeptide and the *nrdB* gene encodes the β-polypeptide. In almost all Eubacteria these two genes form an operon (Figure 2). The eukaryotic RNRs are of the class Ia type. The α– and β–polypeptides in class Ib RNRs are encoded by the *nrdE* and *nrdF* genes and are generally found in an *nrdHIEF* operon (Figure 2), where *nrdH* encodes a physiological reductant (the NrdH-redoxin) that is specific for the class Ib RNRs and *nrdI* encodes a flavodoxin-like protein of unknown function [35, 36]. Class Ib RNRs also differ from class Ia RNRs by lacking the ATP-cone domain and thus have no allosteric overall activity site.

Table 1. Different names for RNR proteins used in the literature and in some public databases

Organism	Class I Ia NrdA	Class I NrdB	Class I Ib NrdE	Class I NrdF	Class II NrdJ	Class III NrdD	Class III NrdG
Escherichia coli	R1, α_2, NrdA, large component	R2, β_2, NrdB, small component	R1E, NrdE	R2F, NrdF		NrdD, α_2, anaerobic RNR	NrdG, β_2, RNR activase
Mycobacterium tuberculosis			NrdE	NrdF1, NrdF2	NrdZ, NrdJ		
Pseudomonas aeruginosa	NrdA	NrdB			NrdJa+NrdJb	NrdD	NrdG
Phage Aeh1	NrdAa+NrdAb	NrdB				NrdD	NrdG
Saccharomyces cerevisiae	RNR1 (NrdA1), RNR3 (NrdA2)	RNR2 (NrdB1), RNR4 (NrdB2)					
Mouse *musculus* constitutive	M1, R1, Rrm1, (NrdA)	M2, R2, Rrm2, (NrdB1)					
Mouse *musculus* p53-inducible		p53R2, Rrm2b, (NrdB2)					
Herpes simplex virus	H1, R1 (NrdA)	H2, R2 (NrdB)					

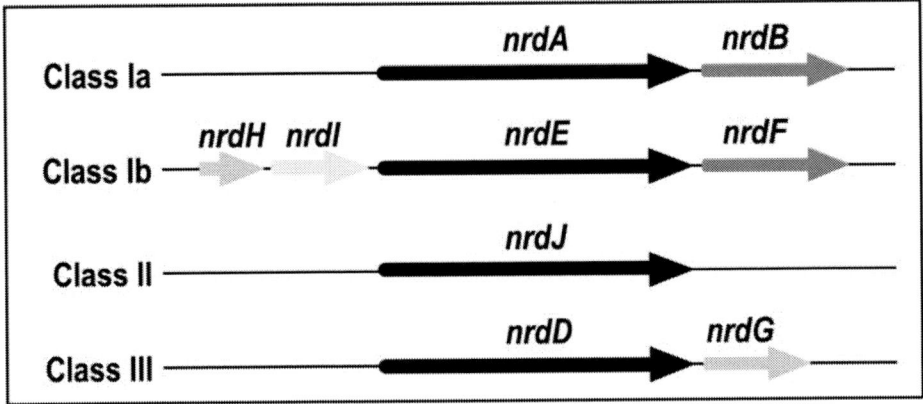

Figure 2. Genomic organization of the different ribonucleotide reductase genes. Catalytic subunits are shown in black and genes with a common evolutionary origin have a similar shade of gray.

Class II RNRs

Class II RNRs are built from a single α–polypeptide encoded by the *nrdJ* gene (Figure 2). The quaternary structure is in most cases either α or α_2 although a α_4 structure has also been reported [25, 37-41].

The 5'-deoxyadenosylcobalamin cofactor binds in the active site region, and it has been shown that the 5'-deoxyadenosyl radical formed after homolytic cleavage of the cofactor is only ca. 6Å from the active site cysteine residue that transiently harbors the thiyl radical [42, 43]. Cleavage of the cobalamin cofactor is oxygen-independent and class II RNRs are often found in facultative Eubacteria and strictly anaerobic Archaea. It has also been found in a few unicellular Eukaryota (see below). Interestingly, some class II RNRs reduce ribonucleoside diphosphates and others reduce ribonucleoside triphosphates. Class II RNRs harbors an allosteric specificity site, but lack the allosteric overall activity site.

Class III RNRs

Class III RNRs are also built from a single α–polypeptide chain, encoded by the *nrdD* gene, with an α_2 quaternary structure [44]. Generation of the stable glycyl radical, located ca. 5Å from the essential active site cysteine that transiently harbors the thiyl radical [24, 45], requires a specific activase encoded by the *nrdG* gene [44, 46, 47] often located in the same operon as the *nrdD* gene (Figure 2). The NrdG protein is a member of the radical-SAM family [48], and contains an iron-sulfur center that can mediate cleavage of *S*-adenosylmethionine to a deoxyadenosyl radical that in turn generates the glycyl radical in NrdD by hydrogen abstraction [49]. The glycyl radical is stable in anaerobic conditions but extremely oxygen-sensitive. Consequently, class III RNRs are confined to Eubacteria and Archaea that grow anaerobically. Class III variants have also been found in a few parasitic eukaryotes (see below). The class III RNRs reduce ribonucleoside triphosphates and usually harbors both an allosteric specificity site and an allosteric overall activity site.

RNR distribution as seen in RNRdb - Ribonucleotide Reductase Database

An open access database of RNRs is available at http://rnrdb.molbio.su.se. RNRdb lists the RNR contents of organisms as deduced from sequencing of genomes, cloned genes, or cDNAs, or from sequencing of isolated polypeptides. All protein sequences (currently more than 3000 entries) are manually curated, and have direct links to their corresponding entries in public database. RNRdb also includes links to the chromosomal location of the gene when known, the three-dimensional structure of the protein if determined, known mutations, and the taxonomy of the organism. Two important features of RNRdb are the possibility to BLAST an unknown protein or nucleotide sequence against the curated and correctly classified RNR protein sequences in the database, and to SEARCH the database for user-defined organisms with user-defined RNR contents. The SEARCH function also allows the output to be restricted to organisms with completely sequenced genomes, to RNRs with known three-dimensional structures and/or to RNRs with known mutations. Future additions to the SEARCH function will include the possibility to restrict searches to RNRs with self-splicing introns and/or self-splicing inteins.

An unexpected observation that is clearly highlighted via RNRdb is that many organisms encode more than one class of RNRs (Table 2). Whereas the majority of the eukaryotic organisms encode solely class Ia RNR, only half of the prokaryotes encode a single RNR operon. A third of the prokaryotes encode two different RNRs, e.g. class I+II, class I+III or class II+III, and have the possibility to grow both aerobically and anaerobically; in theory this possibility is also open to the prokaryotes that encode only a class II RNR. The most enigmatical observation is that 6% of the Eubacteria encode one representative for each RNR class, i.e. have both "belt and suspenders". A few *Firmicutes*, and some α–, β– and γ–*Proteobacteria* are examples of such organisms. Quite often only one of the RNR operons in a prokaryotic organism will follow the classical distribution according to the 16S rRNA phylogeny, whereas other RNR operons are more similar to RNRs in other parts of the 16S rRNA tree. It is thus logic to assume that organisms with more than one class of RNR have acquired additional operons via horizontal gene transfer and that the additional RNRs confer selective advantages to these organisms and therefore have been kept during evolution. Work in the authors' research groups has demonstrated that all three RNR operons encoded by *Pseudomonas aeruginosa* are physiologically important for the organism [50] (Torrents, E. and Sjöberg, B.-M., unpublished).

Table 2. Statistics of RNR combinations in organisms with fully sequenced genomes. Data taken from RNRdb, the Ribonucleotide Reductase database, http://rnrdb.molbio.su.se), in October 2007

Class combination	Archaea	Bacteria	Bacteriophages	Eukaryota	Eukaryotic viruses
Ia	0	101	8	49	92
Ib	0	33	4	0	0
II	13	62	10	0	0
III	10	7	1	0	0
Ia+Ib	0	8	0	0	0
Ia+II	4	38	0	1	0
Ia+III	0	46	12	0	0
Ib+II	0	10	0	0	0
Ib+III	0	55	0	0	0
II+III	12	22	0	0	0
Ia+Ib+II	0	6	0	0	0
Ia+Ib+III	0	23	0	0	0
Ia+II+III	0	27	0	0	0
Ib+II+III	0	4	0	0	0
Ia+Ib+II+III	0	0	0	0	0

With more and more genomes being completely sequenced it has become evident that horizontal gene transfer of RNR genes is not confined to prokaryotic organisms, but is also found among unicellular eukaryotes (*Euglena*,

Dictyostelium, Gibberella, Phytophthora). Interestingly, such occurrences seem to occur preferentially in organisms that are in close contact with bacteria and the class II and III RNRs in these cases are very closely related to prokaryotic counterparts, so the horizontal gene transfer seems to be recent. In at least one case, preliminary studies on *D. discoideum* in the authors' laboratory suggest that both the classical eukaryotic class Ia RNR and a horizontally acquired class II RNR are expressed (Crona, M., Söderbom, F. and Sjöberg, B.-M., unpublished). It is therefore highly plausible that the recently acquired class II and III RNRs give selective advantages to these eukaryotes.

Some fungi and many of the multicellular eukaryotes code for several variants of the class Ia RNR. A well-studied example is the four RNR genes in *Saccharomyces cerevisiae* [51-53], two coding for α–polypeptides (*RNR1* on chromosome V and *RNR3* on chromosome IX) and two for β–polypeptides (*RNR2* on chromosome X and *RNR4* on chromosome VII). The holoenzyme that provides dNTPs for DNA replication consists of a RNR1 dimer plus a RNR2/RNR4 heterodimer, whereas RNR3 is induced upon DNA damage (see below). Other fungi may encode only one gene each for the α– and the β–polypeptides, whereas others may have one α–polypeptide and two β–polypeptides, or vice versa, or the full quadruple set of genes as in the baker's yeast.

Another well-studied example is the mouse that encodes one gene for the α–polypeptide (*Rrm1* on chromosome 7) and at least two genes for β–polypeptides (the conventional *Rrm2* on chromosome 12, and a p53-inducible *Rrm2b* on chromosome 15). The holoenzyme for dNTP synthesis during DNA replication consists of Rrm1 and Rrm2, whereas the gene coding for Rrm2b (also called p53R2) has an upstream region that binds the global tumor suppressor protein p53 and is highly induced when p53 is defective [54-57](see Chapter 5) Occurrence of these three genes seems to be common and is found in all hitherto sequenced mammals, fishes, birds and perhaps nematodes.

RNR genes are also frequently found in dsDNA viruses and bacteriophages with genomes above ca. 100 kb. In eukaryotic viruses they were first observed in the Herpesviridae, and RNRdb currently lists over 100 such viruses with class Ia RNR genes. In Herpesviruses both genes are close together on the chromosome [58, 59], as is also the case in Asfivirus and Baculovirus, but in many other viruses the RNR genes are found at different locations on the viral genome. It has been speculated that the viral RNR offers a selective advantage to the virus, since the viral RNR usually lacks the allosteric overall activity regulation and therefore has no negative feedback on enzyme activity even at very high dNTP levels. A similar hypothesis has also been entertained for the enterobacteriophage T4 [60, 61]. All members of the T-even phage family encode a class Ia RNR operon and a class III RNR operon, and both enzymes lack a functional allosteric overall activity site. All possible RNR classes (Ia, Ib, II and III) are represented among the ca. 50 bacteriophages listed in RNRdb.

Atypical and Unconventional RNRs

With the increasing number of sequenced genomes available in public databases several unexpected variations in the typical RNR classes are also found, and we will describe some of these unconventional RNRs below.

Multiple ATP-Cone RNRs

P. aeruginosa and *Chlamydia trachomatis* have atypically long extensions of the N-terminal part of their class Ia NrdA proteins with almost 220 and 320 extra amino acids, respectively (Figure 3) [27, 62]. Detailed comparative analysis to other well-known NrdA proteins or against the Pfam database [63] revealed that they have duplication or a triplication of the ATP-cone domain in their N-terminal part of the protein. The N-terminal extension has only been studied so far in *P. aeruginosa* NrdA. The wild-type NrdA and two truncated NrdA variants with precise N-terminal deletions were used to study the allosteric properties and the function of these ATP-cone domains. The most N-terminal ATP-cone (ATP-c1) possessed the allosteric function, whereas the internal ATP-cone (ATP-c2) had no allosteric function. The *P. aeruginosa* NrdAB complex has an apparent octameric structure ($\alpha_4\beta_4$) and surprisingly, the first ATP-cone domain has a dramatic effect on the oligomerization since a deletion of this domain produced a tetrameric structure ($\alpha_2\beta_2$), suggesting that the N-terminal domain is crucial for the quaternary structure of the wild-type *P. aeruginosa* NrdAB complex as well for the allosteric regulation of the overall enzyme activity [27].

The duplication in the N-terminal ATP-cone domain as in *P. aeruginosa* NrdA is also found in some members of the β- and γ-proteobacteria and a triplication is found in *Chlamydiaceae*. Surprisingly, a phylogenetic tree based on only the catalytic part of NrdA showed that all these N-terminally extended sequences are clustered in the same branch far away from the corresponding groups of bacteria without extensions. The tree topology suggested that these alterations have occurred recently and frequently. We therefore suggest that the duplications/triplications have occurred by horizontal acquisition of another RNR gene rather than by recombination between chromosomally existing domains. It is a clear illustration of RNR domain plasticity enforcing the modular structures of the different RNR classes (see below).

Split Nrd Genes

It has been taken for granted that the catalytic subunits (NrdA, NrdE, NrdJ, NrdD) of all three RNR classes are encoded by a single gene (*nrdA, nrdE, nrdJ, nrdD*, respectively). This has indeed been the case until the recent characterizations of the class II RNR from *P. aeruginosa* [50] and the class Ia RNR of *Aeromonas hydrophila* phage Aeh1 [64].

P. aeruginosa class II RNR is split and encoded by two consecutive open reading frames denoted *nrdJa* and *nrdJb* separated by 16 bp (Figure 3). This gene structure differs from all hitherto known class II enzymes encoded by a single *nrdJ* gene. The *P. aeruginosa nrdJa* and *nrdJb* genes are linked and transcribed as a single mRNA, and both proteins were expressed. The NrdJa component contains the conserved regions encompassing the activity site and the allosteric specificity site and the NrdJb contains a C-terminal cysteine cluster that interacts with the reducing system. Consequently, the *P. aeruginosa* class II RNR only shows activity when both NrdJa and NrdJb proteins are present. NrdJa by itself lacks enzyme activity even with an artificial reductant or in the presence of wild-type crude extracts containing all the physiological reducing agents. This is the first demonstration of a class II RNR that requires a two-component system for enzyme activity. A comparison with the sequence of *Lactobacillus leichmannii* and the corresponding 3D structure of its complex with a B_{12} analogue, suggest that the B_{12}-binding domain of the *P. aeruginosa* class II enzyme is split between NrdJa and NrdJb. Vitamin B_{12} is a necessity for class II enzymatic activity and at this point, we can merely speculate that the function of the NrdJb might be to sequester and present vitamin B_{12} to NrdJa or that a binding pocket for B_{12} is formed at the protein interface of the NrdJa/NrdJb complex.

The split *nrdJ* gene in *P. aeruginosa* is not unique since this feature is also found in *Azotobacter vinelandii, Magnetococcus* sp., *Methylobacillus flagellatus* and *Photobacterium profundum*. A phylogenetic tree [50] clearly showed a separation of the split NrdJ proteins into a unique and well-defined cluster. Most of the other NrdJ sequences from the proteobacteria are clustered into their corresponding subdivision (α-, β-, and δ-proteobacteria) but the cluster of bacteria with split NrdJ proteins is more diverse; *P. aeruginosa*, *A. vinelandii* and *P. profundum* belong to the γ-proteobacteria, *M. flagellatus* is classified as a β-proteobacterium and *Magnetococcus* sp. is an unclassified proteobacterium.

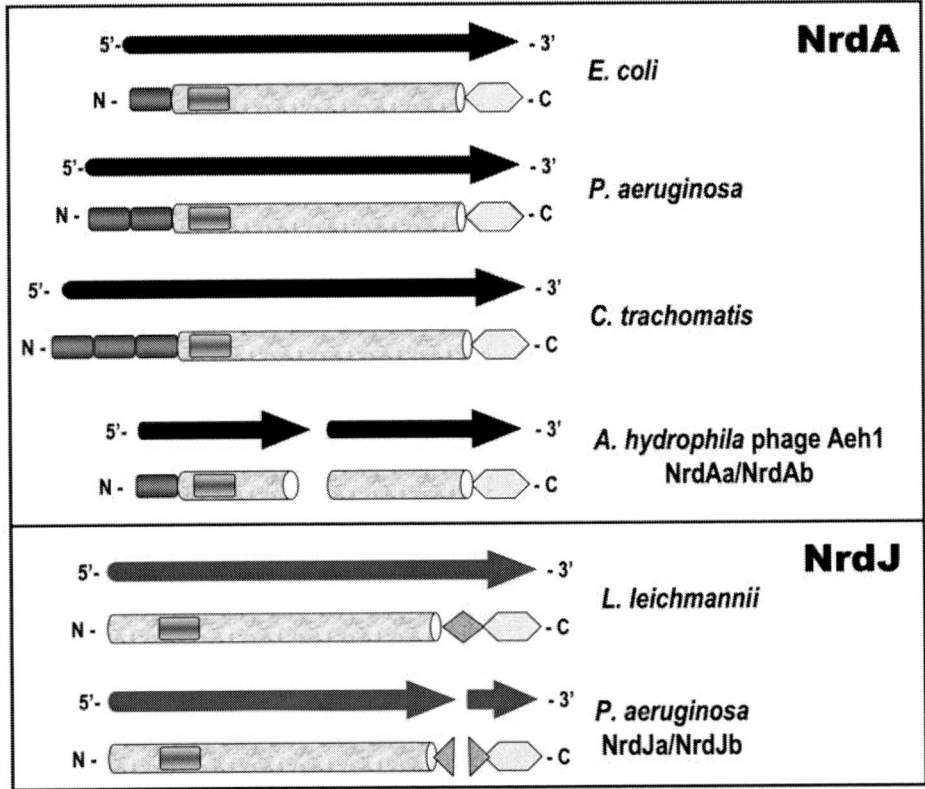

Figure 3. Schematic representation of atypical class I and II RNR sequences. Black arrows represent gene structures and the protein domains are shown below (dark gray box, ATP-cone; light gray cylinder, catalytic core with allosteric specificity in medium gray; gray hexagon, C-terminal redox-active cysteine domain; medium gray diamond (split for *P. aeruginosa*), B_{12}-binding domain). The upper part of the figure shows the normal *E. coli* prototype NrdA protein composition and *P. aeruginosa* NrdA with an ATP-cone duplication and *C. trachomatis* with an ATP-cone triplication at their N-terminal parts (dark boxes). The split *nrdA* gene and the resulting two proteins NrdAa and NrdAb from *A. hydrophila* phage Aeh1 are shown. In the lower part of the figure the split NrdJa+NrdJb protein from *P. aerugionosa* is compared to the *L. leichmannii* full-length NrdJ protein.

The class I RNR in *A. hydrophila* phage Aeh1 is even more perplexing because its *nrdA* gene is split in two by the insertion of a homing endonuclease gene (*mobE*) (Figure 3). The insertion creates a genes-in-pieces arrangement, where the NrdA function is encoded by two independent genes, *nrdAa* and *nrdAb* [64]. Significantly, the Aeh1 *nrdA* gene is split such that *nrdAa* encodes some of the active site residues and *nrdAb* encode the other active site residues. It is amazing that phage Aeh1, despite the *mobE* insertion, can produce a functional class I RNR with a composite active site formed by residues from the two NrdA pieces in complex with the NrdB protein. This finding raise the intriguing question by which mechanisms active site residues from two different proteins are assembled correctly to form a fully functional active enzyme with similar specific activity to other characterized class Ia RNR systems.

NrdD Metal-Binding Motifs

Deduced amino acid sequences for the approximately 250 class III RNRs in the RNRdb show that the C-terminal metal binding motif in NrdD can be of two different types, the predominant $CXXCX_{14}CXXC$ motif with the site of the glycyl radical 15 residues further to the C-terminal end [65], and another variant with a $CXXHX_9CXXC$ motif found in ca. 50 sequenced *nrdD* genes (Figure 4; Raleiras, P. and Sjöberg, B.-M., unpublished). The structure of the metal binding motif is reminiscent of rubredoxin (see Chapter 9). Interestingly, the mixed Cys/His-containing motif that is found e.g. in *P. aeruginosa* NrdD has a potential glycyl radical harboring motif within the rubredoxin-like domain at Gly644 adjacent to the Cys640-His643 part of the motif (Figure 4). Several other Proteobacteria, but far from all, have this type of NrdD. About half of the mixed Cys/His-containing NrdDs encode both a potential glycyl radical-harboring residue adjacent to the Cys/His motif as in *P. aeruginosa* NrdD and another potential glycyl radical site positioned either 11 or 16 residues C-terminal of the Cys/Cys motif; among these are some Firmicutes, Bacteriodetes and a few Euryarchaea (Figure 4). Current work in the authors' laboratory aim at mapping the glycyl radical site in these two variants of NrdD proteins with the mixed Cys/His motif (Raleiras, P. and Sjöberg, B.-M., unpublished).

An anaerobically DNA-deficient phenotype in *Ralstonia eutropha* (a β–Proteobacterium) was mapped as G650A corresponding to the position of Gly644 in *P. aeruginosa* NrdD (marked with an arrowhead at the bottom in figure 4), suggesting that Gly650 is the site of the glycyl radical in this type of NrdD [66]. The *R. eutropha* class III RNR is encoded on a megaplasmid [67] and hence may not be the *R. eutropha* "parental" RNR, whereas class I and II RNRs are encoded on the *R. eutropha* chromosome I [68]. Similarly, subgroups of class I and II RNRs are also obvious from amino acid sequence comparisons, but in absence of additional biochemical or genetic experiments showing that the subgroups form a distinct subclass we would suggest caution in making and naming novel subclasses of RNR.

```
              540                                  570
               |    *    *                       *   *         |              v
PhT4NrdD    - NM PVD KCFTCGSTH EMT PTE NGFVCSI CGET DPK KMN TIR RTC GYL GNP -
EcolNrdD    - NT PID ECYECGFTG EFE CTS KGFTCPK CGNH DAS RVS VTR RVC GYL GSP -
CperNrdD2   - SP LNR YCEEHGYV - - - - - KE RVEECPI CGKK - - - - LD LYQ RIT GYL RKV -
CtheNrdD    - TP TFS ICKDHGYI - - - - - KG EHFKCPE CGND - - - - TE VWS RVT GYL RPV -
ReutNrdD    - TP TFS ICPTHGYL - - - - - AG EHPFCPR CDEE ILA RKR DQL AA*
PaerNrdD    - TP TFS ICPRHGYL - - - - - AG EHEFCPK CDEE LLQ RST PKS CCG GCG G*
               |          ^                                       |
              640                                                660
```

Figure 4. Variants of the metal binding motif in class III RNRs. Two representatives of each variant is shown; PhT4NrdD, bacteriophage T4 NrdD; EcolNrdD, *E. coli* NrdD; CperNrdD2, the second NrdD of *C. perfringens*; CtheNrdD, *C. thermocellum* NrdD; ReutNrdD, *R. eutropha* NrdD; PaerNrdD, *P. aeruginosa* NrdD. Asterisks mark iron ligands, and arrowheads indicate known or potential glycyl radical positions. Numbering refer to bacteriophage T4 (top) and *P. aeruginosa* (bottom).

Class I RNR Sequences Lacking a Stable Tyrosyl Radical

C. trachomatis is an abundant human pathogen and the most common sexually transmitted bacterial disease in the developed world. *C. trachomatis* encodes a class Ia RNR with an NrdB protein that constitutes a fascinating exception to most other NrdBs. Despite the fact that the *Chlamydia* NrdB protein has a significant overall homology to other NrdB proteins it has a phenylalanine (Phe127) in the position equivalent to where other NrdB proteins have the tyrosine that carries the stable radical [62, 69]. *C. trachomatis* cannot import deoxyribonucleotides from its environment and thus has to synthesize its own deoxyribonucleotides via its unique class Ia RNR. Even more surprising is the fact that the combination of pure NrdA and NrdB proteins leads to an active holoenzyme even though no traces of a tyrosyl radical signal have ever been seen in EPR. As will be described in Chapter 6, the *C. trachomatis* enzyme has a high-valent diiron (Fe^{III}-Fe^{IV}) site instead of the tyrosyl radical [70].

This unconventional NrdB sequence has been seen not only in *C. trachomatis*, but also in all Chlamydia sequences and in a few other microorganisms. A note of caution is that although several NrdB-like proteins have this motif only few of them are expected to have functional relevance because some of these organisms seem to lack the other catalytic subunit, the NrdA protein. E.g. *Mycobacterium* and *Sulfolobus* species encode such an NrdB-like protein but no traces of an NrdA protein. In contrast species like *Halobacterium salinarum*, *Geobacillus kaustophilus* and *Chloroflexus aurantiacus* may be similar to *C. trachomatis* as their genomes encode both the NrdB-like protein and an NrdA protein (see Chapter 6).

```
                  *          100                 120  * *     ♦           140                        180
     SagaNrdF  TGLTLLDSVQATVGDIAQIKHSQTDHEQVIYANFAFMVAIHARSYGTIFSTLCTSQQIEEAHEWVVDTESLQARSRILIPFYT--  160
     SpyoNrdF  TGLTLLDTVQATVGDVAQIQHSQTDHEQVIYTNFAFMVGIHARSYGTIFSTLCSSEQIEEAHEWVVSTQSLQDRSRVLIPYYT--  161
     MgenNrdF  TGLTLLDTIQATIGDICQIDHALTDHEQVIYANFAFMVGVHARSYGTIFSTLCTSEQINAAHEWVVNTESLQKRAKALIPYYT--  164
     EcolNrdF  TGLTLLDTLQNVIGAPSLMPDALTPHEEAVLSNISFMEAVHARSYSSIFSTLCQTKDVDAAYAWSEENAPLQRKAQIIQQHYR--  143
     EcolNrdB  KYQTLLDSIQGRSPNVALLPLISIPELETWVETWAFSETIHSRSYTHIIRNIVNDPSV--VFDDIVTNEQIQKRAEGISSYYDEL  161
     Conservation

                       180           200                 220                   240   * *
     SagaNrdF  ---------GDD--------PLKSKVAAAMM-----PGFLLYGGFYLPFYLSARGKLPNTSDIIRLILRDKVIHNYYS  216
     SpyoNrdF  ---------GDD--------PLKSKVAAAMM-----PGFLLYGGFYLPFYLSARGKMPNTSDIIRLILRDKVIHNYYS  217
     MgenNrdF  ---------GND--------PLKSKVAAALM-----PGFLLYGGFYLPFYLSSRKQLPNTSDIIRLILRDKVIHNYYS  220
     EcolNrdF  ---------GDD--------PLKKKIASVFL-----ESFLFYSGFWLPMYFSSRGKLTNTADLIRLIIRDEAVHGYYI  199
     EcolNrdB  IEMTSYWHLLGEGTHTVNGKTVTVSLRELKKKLYLCLMSVNALEAIRFYVSFACSFAFAERELMEGNAKIIRLIARDEALH--LT  244
     Conservation
```

Figure 5. Sequence alignment of the non-functional NrdF sequences. Non-functional NrdFs from *S. agalactiae* (SagaNrdF), *S. pyogenes* (SpyoNrdF) and *M. genitallium* (MgenNrdF) are compared to the functional NrdF from *E. coli* (EcolNrdF) and NrdB from *E. coli* (EcolNrdB). The tyrosyl radical position is shown by a diamond, and the six iron ligands in *E. coli* NrdB/F are indicated by stars. Numbering for each protein sequence is shown to the right. The degree of conservation is indicated below the alignment.

Non-Functional RNR Subunits

In class I RNRs, the tyrosyl radical is generated in the NrdB/NrdF proteins by the oxygen-dependent oxidation of the diiron-carboxylate center coordinated by six highly conserved residues (see Chapter 6) present in almost all known NrdB/NrdF proteins with the exception of few remarkable cases [33, 53, 69, 71-73]. In all *Mycoplasma* species and in *Streptococcus pyogenes*, *Streptococcus agalactiae* and *Streptococcus equi* the NrdF protein has an amino acid substitution in three out of the six important residues involved in iron binding (Figure 5). It is important to note that the gene encoding this unusual NrdF protein is highly conserved among *Mycoplasma/Streptococcus* and arranged in a different gene order (*nrdFIE*) compared to the gene order of the common class Ib components (*nrdHIEF*). In *Streptococcus* the unconventional *nrdF* locus is not the only RNR encoded in their genomes since they also have a set of the common class Ib genes (*nrdHEF*) and a class III RNR set of genes (*nrdDG*). However, the unconventional *nrdFIE* locus is the only RNR class encoded in the genomes of *Mycoplasmas*. A global transposon mutagenesis of *M. genitalium* gave a viable transposon insertion in the *nrdE* gene [74], indicating that these intracellular parasites possibly encode a non-functional RNR. Unpublished data also showed that the purified *S. pyogenes* NrdF protein that lacks the common iron ligand residues has no iron bound, no tyrosyl radical and no enzyme activity in complex with either of the NrdE proteins (Roca, I., personal communication). Several question come to mind. Why do these microorganisms keep the highly conserved genes in their genomes? Have we really found the correct conditions to assay for enzyme activity in these proteins? Further work is needed to elucidate the role of this set of genes.

Chapter III

RNR Diversification

The three separate classes of RNRs known in nature that differ in cofactor requirements, quaternary architectures and behavior towards oxygen (see above) also have striking similarities. They share similar modes of protein regulation and reaction mechanism and it is generally accepted that the three RNR classes have a common evolutionary origin [14, 25, 75, 76]. RNR evolved early in the history of life on Earth, under anaerobic conditions during the transition of an RNA to a DNA world. RNR is considered to be one of the oldest enzymes and a component of the very first organisms that appeared more than 3.5 billion years ago [77, 78]. Regarding the origin of the three classes of RNR, there is an agreement on a divergent evolution of the classes with class I RNR being more recent than the class II and III. Class I RNR is oxygen dependent and hence could not have functioned under early anaerobic conditions. However the intriguing question which of class II and III RNRs has most resemblance to the "ur-reductase" is still unresolved [75, 76]. Several reviews and articles have sought to answer the question using very different approaches [75-80]. In this review we will not focus on that topic but instead consider the essential role that modularity has played for the specification of three different RNR classes.

Modularity Hypothesis for RNR Class Diversification

All three RNR classes have in common the catalytic core where the active site of the enzyme and its specificity regulation are located. We suggest that protein modularity built on this common catalytic core plays an important role in class diversification within the RNR family. The modularity concept is an attractive hypothesis for how proteins may have evolved and diversified in nature [81, 82] and could easily explain why and how the three contemporary RNR classes reduce ribonucleotides using essentially the same reaction mechanism. It is interesting to note that the central catalytic core (the 10-stranded β/α-barrel structure) has been highly constrained during evolution as seen in the conserved three-dimensional structure [23-25, 83], whereas the different domains connected to the core region are more labile but essential and important to define each RNR class. Which are the different modules that created the three contemporary RNR classes? Domains fused to the C-terminal end (or separate proteins interacting with the C-terminal end) of the core are mainly involved in enabling and fine-tuning the reaction mechanism, whereas domain(s) fused to the N-terminal end are involved in regulation of the enzyme activity (Figure 6).

To start with the class-specific C-terminal additions that enable catalysis, the class II RNR requires a 5'-deoxyadenosylcobalamin (AdoCbl or vitamin B_{12}) cofactor. Only regions but not individual side chains of the class II structure that interact with an AdoCbl analogue were identified in the structure of the *L. leichmannii* RNR [25], whereas the *T. maritima* RNR was crystallized with a C-terminal deletion and in the absence of AdoCbl [37]. It is obvious that at least part of what makes up the AdoCbl-binding domain is in the C-terminal part of the class II enzyme, suggesting that the acquisition of B_{12}-binding occurred by addition of a C-terminal domain in class II RNR.

Class III has acquired a completely different C-terminal domain that has the capacity to harbor the stable glycyl radical [24]. Apart from the glycine residue this addition also comprises the rubredoxin-like domain with four metal-binding side chains [65] as described above (cf. figure 4). This domain is also present in the anaerobic enzyme pyruvate formate lyase (PFL) and close relatives [84] and constitutes a signature for all class III RNRs.

Class I is a special case and needs a separate protein, the NrdB/F with the capacity to harbor the stable tyrosyl radical. It is clear that NrdB/F mechanistically interacts with a radical transferring tyrosine pair (Tyr730 and Tyr731 in *E. coli* NrdA) that is located in the C-terminal part of NrdA/E. It was in fact pointed out that the AdoCbl derivative in class II RNR occupies an equivalent position in the structure as the tyrosine pair in class I RNR [25].

In addition to these class-specific domains/protein, all three enzymes also need additional specific domains and/or proteins for turnover and catalysis. The last 14-24 residues at the C-terminal end of class I and II RNRs are flexible and not resolved in the crystal structures [23, 25, 85]. This part contains at least one redox-active cysteine pair (in many class II enzymes a cluster of several cysteines) that interacts with and accepts reducing electrons from different redoxins proteins (thioredoxin, glutaredoxin, NrdH-redoxin [36, 86]; not shown in figure 6). As part of the catalytic turnover mechanism the reduced C-terminal cysteine pair interacts and reduces the active site disulfide that is formed during catalysis, and the redoxin protein then reduces the oxidized C-terminal disulfide. This maintains efficient catalytic turnover of class I and II RNRs.

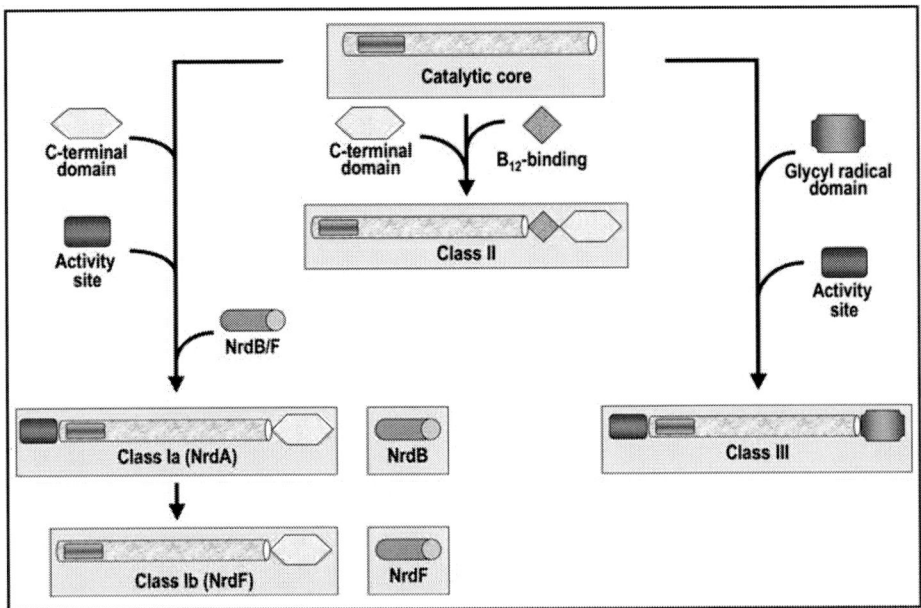

Figure 6. Schematic representation of the modularity hypothesis for RNR class diversification. The catalytic core is represented by a gray cylinder with the allosteric specificity indicated in medium gray. Activity site indicates the ATP-cone domain, and C-terminal domain indicates redox-active cysteines. The function of other domains and proteins are indicated directly in the figure.

Class III RNRs on the other hand need a specific RNR activase (NrdG; not shown in figure 6) to generate the glycyl radical in the C-terminal domain. The RNR activase is a member of the abundant radical-SAM protein family [48].

If we then turn to the N-terminal end, class Ia, III and some class II RNRs have an ATP-cone [34], the common allosteric domain that in the RNRs define the overall activity of the enzyme. It binds either ATP or dATP as allosteric effectors and has hence the possibility to monitor the level of reduced deoxyribonucleotides in the cell at any given time. Class Ib RNRs lack the ATP-cone and we suggest that class Ib evolved from class Ia by deletion of the ATP-cone.

It is evident that evolution of contemporary RNR classes involved subtle shuffling and recombinations of different domains (ATP-cone domain, radical-harboring/cofactor-binding domains or proteins, etc) onto a central catalytic domain and in this way shaped the three well-defined RNR classes that we know of today.

Chapter IV

Regulation of RNRs

Ribonucleotide reduction plays an important role in the cellular metabolism because it provides a sufficient pool of balanced dNTPs required for DNA synthesis and repair. An alteration of the dNTP concentrations would lead to an increased mutation frequency, which could be fatal for the organism's viability. The fine-tuning of dNTP levels is achieved by at least two different regulatory mechanisms, transcriptional regulation of the *nrd* genes and allosteric control of the RNR enzyme. The allosteric regulation has been described above. Here we will discuss the organization of RNR genes and current knowledge about their genetic regulation in bacteria, and the regulation of RNR genes in bakers' yeast. Regulation of mammalian RNRs will be described in Chapter 4.

Archaea

In general the transcriptional regulation of the RNRs in prokaryotes is complicated because they are able to produce more than one type of RNR simultaneously. However, except for the global transcriptional analysis in *Halobacterium* sp. NRC-1 under UV-radiation, *nrd* transcriptional regulation studied in Archaea is lacking. In *Halobacterium* increased class II RNR transcription was found after 1-3 hours of UV-radiation [87] demonstrating that *nrd* expression and DNA repair mechanisms are tightly linked.

Eubacteria

Transcription of the different sets of *nrd* genes in bacteria responds to changes in the environmental cues. This enables simultaneous expression of several RNRs in organisms that encode more than one RNR class. Usually one "predominant" RNR is responsible for support of bacterial growth and is regulated depending on the cell cycle to satisfy the changing demands for dNTP synthesis. Other RNR classes that are present in the same genome respond to other regulatory mechanisms that differ from those controlling the "predominant" RNR class.

Bacterial RNR Operon Structures

Normally the *nrdA* and *nrdB* genes encoding class Ia RNR in bacteria constitute a tightly regulated transcriptional unit (Figure 2). In a few cases, like in *Helicobacter pylori,* the two genes are not forming an operon and *nrdA* is not even close to the *nrdB* gene.

The class Ib RNR operon contains four different genes, where *nrdE* and *nrdF* encode the class Ib RNR proper, *nrdH* encodes the specific NrdH-redoxin, and *nrdI* encodes a flavodoxin-like protein of unknown function. The order of genes in the class Ib operon is typically *nrdHIEF* (Figure 2) but different variations have been found and class Ib operons have the highest diversity of all RNRs. For example in most of the Gram-positive low GC content bacteria the *nrdH* is not part of the operon but located elsewhere on the chromosome. In some members of the *Mycobacteriaceae* family (*Mycobacterium tuberculosis*, *Corynebacterium diphtheriae*, *Corynebacterium ammoniagenes* and *Corynebacterium glutamicum*) the intergenic region between *nrdE* and *nrdF* is of variable length and encodes a transcriptional factor of the GntR (gluconate regulator) family [88]. In this case the *nrdHIE* normally form an operon independently from the *nrdF*. In the *Mycoplasma* species the gene order is *nrdFIE*, and as described above the deduced NrdF protein lacks three out of the six iron ligand residues (Figure 5). Some species have a duplication of the class Ib operon, and the two sets in *S. pyogenes* and *S. agalactiae* form one *nrdIEF* operon and one *nrdFIE* operon similar to the one in *Mycoplasma* species [89].

A single *nrdJ* gene normally encodes the class II RNR (Figure 2). Some bacterial species encode two (*Rhodobacter sphaeroides* and *Paracoccus denitrificans*) and as described above some γ, β–, and unclassified proteobacteria have a split *nrdJ* (*nrdJa* + *nrdJb*, cf. Figure 3).

Two different genes, *nrdD* coding for the RNR proper and *nrdG* coding for the RNR activase encode class III RNR components. Normally the two genes are arranged in an operon (Figure 2) and some bacteria even contain two sets of *nrdDG* genes like in *Bacillus thuringiensis* serovar *israeliensis* and in *Clostridium perfringens*. Typically in Archaea the *nrdD* and *nrdG* genes do not form an operon.

RNR Regulation in Enterobacteriaceae

The most extensively studied transcriptional regulation of RNR genes is in *E. coli*. This organism and the entire *Enterobacteriaceae* family [90] contain genes for two aerobic RNRs (class Ia and Ib) and one anaerobic RNR (class III). In 1984 Carlson and Fuchs cloned the first RNR operon corresponding to the *nrdAB* genes from *E. coli* [6]. Almost a decade later, the *nrdD* and *nrdG* genes for the anaerobic RNR in *E. coli* were cloned and sequenced [47, 91, 92]. At about the same time, the third RNR in *Enterobacteriaceae*, the class Ib, was discovered and sequenced, first in *Salmonella typhimurium* [93] and later the *E. coli* sequence became publicly available via the genome sequencing project for this organism [94]. Several groups have studied the transcriptional regulation of these three operons in *E. coli* but it is not yet well understood due to the many factors involved in their regulation. Here we will describe the current knowledge.

Fuchs and co-workers did an exhaustive analysis of the promoter region of the *nrdAB* genes in *E. coli* and identified several promoter regions important for the cell cycle regulation and transcription of the genes [92, 95]. A summary of the upstream region of the *E. coli nrdAB* genes is presented in figure 7. Several years before cloning of the *nrdAB* operon they observed that inhibition of DNA synthesis by chemical inhibitors, thymidine starvation, and non-permissive conditions for temperature-sensitive mutants in DNA elongation (*dnaE* and *dnaB*) or initiation (*dnaA*) led to an increase of class Ia RNR [96]. In addition, they first described that mutations in the *lexA*, *recA*, *recB* and *recC* regulators result in increased synthesis of RNR but did not identify the direct link of these transcriptional regulators to *nrdAB* transcription. Fuchs and co-workers finally localized two 9-bp DnaA boxes upstream of the -35 region of the promoter that confer a cooperative positive effect on the *nrdAB* transcription when DnaA is bound [92]. The Fis (<u>f</u>actor for <u>i</u>nversion <u>s</u>timulation) protein was also found to have a positive effect on transcription correlating the importance of DNA supercoiling to the regulation of the *nrd* genes [92]. That the *nrdA* transcription is affected by the DNA topology was recently corroborated in a survey of the genomic transcriptional response of *E. coli* to loss of chromosomal supercoiling [97, 98]. Fuchs and co-workers had established that neither Fis nor DnaA was required for *nrd* cell cycle regulation and later identified a *cis*-acting upstream AT rich sequence (at -132 bp from the transcriptional start point) and a 45-bp inverted repeat (centered at -93 bp) required for cell cycle regulation [95]. Other studies suggested that the IciA protein functions as a transcriptional activator for the *nrdAB* genes by binding to the AT-rich upstream region of the *nrd* promoter [99].

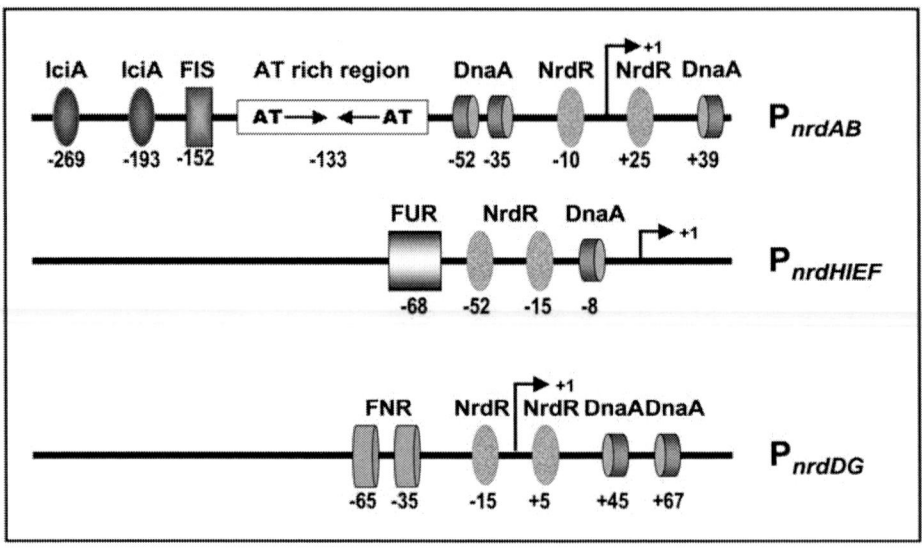

Figure 7. Scheme of the three *E. coli nrd* promoter regions. The promoter regions of *nrdA*, *nrdE* and *nrdD* are shown with possible regulatory regions and the position of each box is shown below. The arrow and +1 represent the transcriptional start points.

However recent evidence established that the DnaA protein interacts with the *nrdAB* promoter and stimulates *nrdAB* expression via the two identified DnaA boxes, and led to the proposal of a complex regulatory mechanism involving the initiation of DNA replication and dNTPs synthesis [100]. The nucleotide-loaded state of DnaA would regulate the transcription of the *nrdAB* genes. The conversion of ATP-DnaA to ADP-DnaA after initiation of DNA replication would allow increased expression of the *nrdAB* genes and consequently, during the cell cycle there would be a coordinated *nrdAB* expression and replication initiation. Additional work demonstrated that the *nrdAB* genes were induced also when both thioredoxin 1 (*trxA*) and glutaredoxin 1 (*grxA*) were mutated simultaneously demonstrating that the class Ia system is highly dependent on functioning reduction systems [101]. A triple mutation in thioredoxin 1 and 2 and glutaredoxin 1 was not viable suggesting that such cells cannot reduce the essential class Ia RNR enzyme [102]. Any conditions that trigger the SOS system, like DNA damage or DNA replication inhibition also induce the transcription of the *nrdAB* and *nrdEF* genes, but the response is completely independent of the SOS response and consequently not dependent on the LexA regulator [103, 104]. The link coupling DNA synthesis inhibition to regulation of class Ia is still unknown and further experiments should be done to elucidate this question.

Class Ib was initially described as a dormant gene because no transcription was observed in normal cells grown under standard laboratory conditions and because it did not support growth of an *nrdA* temperature sensitive mutant at restrictive temperature. However, increased transcription of both class Ia and Ib was observed when cultures were treated with the radical scavenger hydroxyurea (HU) or chemicals that inhibit DNA synthesis [105]. With the increased availability of high-throughput methods and bioinformatics we can change our minds and adopt an unbiased view to what the role of class Ib is. Several studies, mainly microarray data, highlight for the first time a possible function of the class Ib genes under special growth conditions. The expression of the *nrdHIEF* genes was triggered in response to several different compounds generating oxidative stress, but the mechanisms by which the class Ib *nrd* genes were expressed remained unknown. Participation of several major global regulators (RpoS, Fis, cAMP, OxyR, SoxR/S, RecA) were excluded [106, 107]. In addition, the *nrdHIEF* genes were induced in *E. coli* lacking Trx1 and Grx1, the two main redoxins for the correct function of the NrdAB system [106]. Other studies presented the induction of these genes by different reactive nitrogen species (acidified sodium nitrite, nitrosylated glutathione, or the nitric oxide generator GSNO) but without identifying the specific transcriptional factor or the possible pathways involved [108-110]. A recent observation suggests a possible function for the class Ib RNR in *Enterobacteriaceae* under iron limitation. A Fur (ferric uptake regulator) box was described in the promoter region of the *nrdHIEF* [111] and global transcriptional analysis of an *E. coli fur* mutant showed that the *nrdHIEF* genes were derepressed [112]. Similar results were also observed in *Yersinia pestis* in which *nrdHIEF* was identified as a set of genes that were negatively regulated by Fur [113]. A Fur-consensus box seems to be present also in the *nrdHIEF* promoter region in *S. typhimurium* (Sala, I., personal communication) suggesting a common regulatory mechanism in all *Enterobacteriaceae*. A summary of the upstream region of the *E. coli nrdHIEF* genes is presented in figure 7.

Class III RNR genes (*nrdDG*) are strongly expressed under anaerobic conditions and are completely dependent on the transcriptional activator FNR (fumarate and nitrate reduction) but independent of the two-component systems ArcAB (aerobic respiration control). The transcriptional start point was described to be at 154 nucleotides upstream of the protein start codon. One FNR box was suggested in the *nrdD* sequencing work [91] but recently two FNR boxes were located around -35 (FNR-1) and -65 (FNR-2). They have significant similarity to the consensus FNR-binding sequence and were described to work coordinately for a tight transcriptional regulation of the *nrdDG* genes in the transition from aerobic to anaerobic environments [114]. Knock-out mutants of either *nrdD* or *nrdG* cannot grow during strict anaerobiosis but surprisingly these mutants grow well under microaerophilic conditions by overproducing the aerobic class Ia enzyme [115]. A summary of the upstream region of the *E. coli nrdDG* genes is presented in figure 7.

The class Ib genes are completely inactive during anaerobiosis and the promoter is insensitive to oxygen availability. Instead, the class Ia genes (*nrdAB*) are slightly inhibited during prolonged growth at anaerobic conditions [114, 116] and a global analysis using a FNR mutant showed a reduction of transcription in this genomic background [117]. But what is the role of each RNR class in *E. coli* (*Enterobacteriaceae)*? It is evident that class Ia RNR is the "predominant" RNR during standard laboratory condition. Class Ia is distinctly cell cycle regulated and responds rapidly to any DNA damage. Class III is actively transcribed under anaerobic conditions via activation by the FNR regulator and it is essential during anaerobiosis [115]. Perhaps it is now timely to also investigate the function of class Ib. Current experiments highlight two main conditions in which class Ib is highly induced compared to class Ia RNR genes, oxidative/nitrosative stress and iron limitation. These conditions resemble the conditions that a bacterium encounters during the infectious process. A global transcriptome analysis of *Salmonella enterica* within macrophages revealed compelling support for such a hypothesis. Class Ia genes were found to be downregulated during the infection as opposed to class Ib genes that were highly expressed during infection [118]. The class III genes showed a weak induction at later stages of infection. These results suggest that class Ib is the RNR that is highly induced under the infectious process and the presence of a Fur box located in the promoter region of the *nrdHIEF* operon in *Enterobacteriaceae* supports this hypothesis because the Fur protein is known to regulate several virulence-related genes. A summary of factors that stimulate and repress the different *nrd* genes in bacteria is shown in figure 8.

RNR Regulation in Other Gram-Negative Bacteria

Extensive RNR expression studies have been performed with *P. aeruginosa*. This bacterium is one of few bacteria that encode genes for all three RNR classes (Ia, II, III), and as described above its class II RNR is split and encoded by two adjacent genes, *nrdJa* and *nrdJb* [50]. Both class Ia and class II enzyme activities were detected in normal growing cells [119].

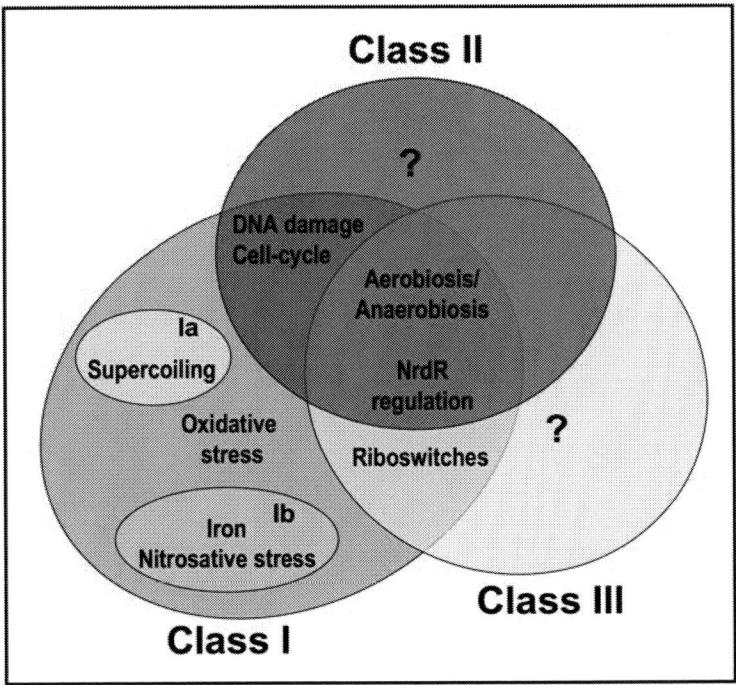

Figure 8. Summary of the different environmental cues that induce or repress the three *nrd* genes in *Enterobacteria*. Specific class Ia and Ib gene regulation is shown in small circles inside the class I diagram. Specific factors that regulate only class II or III are not known and are represented by a question mark.

Quantitative analysis by real-time PCR showed that class Ia and II are growth phase regulated but with opposite expression patterns. Class Ia expression, which is the "predominant" class, reached its highest expression at the beginning of the exponential phase and decreased at the beginning of the stationary phase. Class II transcription instead followed an inverted pattern with the highest transcriptional activity at the stationary phase [50]. Interestingly class II RNR was only able to support class Ia deficiency during aerobic conditions and when the specific cofactor AdoCbl was added to the medium.

These results showed a strict dependence of class II enzyme activity on the presence of vitamin B_{12}. As expected the *nrd* genes of *P. aeruginosa* were induced when DNA replication is inhibited, in this case with HU. Surprisingly the increased transcription observed with the class II genes was 15-fold whereas the class I gene expression was increased only two-fold. These findings suggested that class II RNR is primarily used for DNA repair and/or possibly DNA replication at low oxygen tension [50]. In the related bacterium, *Pseudomonas stutzeri* the expression of the *nrdD* gene appears to be controlled by an FNR-type regulator [120]. This can also be the case for the *P. aeruginosa nrdD* but this hypothesis needs to be tested.

Data not published but available in the authors' laboratories showed that the *P. aeruginosa* class II and III genes are important during the infectious process, using *Drosophila melanogaster* as a model system. There is a distinct shift in *nrd* expression during infection with down-regulation of class Ia and up-regulation of class II and III. In addition, flies had an increased survival after infection by *P. aeruginosa* class II and III knockout mutants compared to wild-type bacteria. These results suggest a tight differential regulation of the three types of RNR classes present in *P. aeruginosa* during the infectious process (Torrents, E. and Sjöberg, B.-M., unpublished).

RNR Regulation in Gram-Positive Bacteria

The transcription of the different *nrd* genes have been studied in several species from the Actinobacteria and among them *C. ammoniagenes, M. tuberculosis, Staphylococcus aureus, Streptomyces coelicolor,* and in addition *Bacillus subtilis* a member of the Firmicutes. Like for other bacterial families the RNR class distribution among the Actinobacteria and Firmicutes is species-specific and each organism encodes a specific combination of RNR classes. *C. ammoniagenes* encodes only class Ib RNR, *M. tuberculosis* encodes class Ib and II, *S. aureus* encodes class Ib and III, *S. coelicolor* encodes class Ia and II, and *B. subtilis* encodes two sets of class Ib RNRs with one copy in a prophage region and this RNR seems to be non-functional [121, 122]. Interestingly, the other class Ib copy is essential for both aerobic and anaerobic growth of *B. subtilis* [123].

C. ammoniagenes and *M. tuberculosis* share the order of class Ib genes and the *nrdHIE* genes form a unique transcriptional unit independent of the *nrdF* gene. The intergenic region between the two transcriptional units is 1171 bp for *C.*

ammoniagenes and 2707 bp for *M. tuberculosis* and the two mRNAs are transcribed from separate promoters with common regulatory mechanisms such that the same environmental conditions affect both transcriptional units at the same time [89].

In *C. ammoniagenes* both transcriptional units (*nrdHIE* and *nrdF*) are cell cycle regulated and follow the characteristic pattern of being highly induced at the beginning of the exponential phase when high dNTP pools are needed [89]. Oxidative stress (H_2O_2) induced the expression of these genes, as for class Ib genes in other organisms, but the link between oxidative stress and induction of transcription was not identified. A DtxR box (analogous to the Fur box in Gram-positive bacteria) was postulated in the promoter regions and the addition of manganese highly induced expression of both transcriptional units [89]. Later, a DtxR box was described for the promoter regions of the related *C. diphtheriae nrdHIE* and *nrdF* genes [124], supporting the results in *C. ammoniagenes* and suggesting that class Ib RNR is important under iron limitation as discussed above.

In *M. tuberculosis* with both class Ib and II RNRs, the class Ib encoded by the *nrdHIE* and *nrdF2* loci is the "predominant" RNR that supports the growth of this bacterium [125]. Previously an extra open reading frame was named *nrdF1* (Rv1981) [125], but there is no adjacent *nrdE* gene, and the NrdF1 protein shows no enzyme activity in presence of the *M. tuberculosis* NrdE protein (Högbom, M., personal communication) that is encoded in the *nrdHIE* operon three genes upstream of the functional *nrdF2* operon. The *M. tuberculosis nrdJ* (class II RNR) is not able to complement a class Ib RNR deficiency [125]. Transcriptional analyses showed that the *nrdJ* gene was expressed under oxygen limitation where the *nrdHIE*/*nrdF2* genes were repressed. A priori we expect *nrdJ* to be important under anaerobic conditions but a *nrdJ* knock-out mutant was viable *in vitro* in microaerophilic/anaerobic conditions and in lungs of B6D2/F1 mice [126]. Theses results suggest that *M. tuberculosis* infectivity apparently depends on the activity of class Ib RNR.

S. aureus relies on the class Ib genes for aerobic growth and under anaerobic conditions the *nrdDG* genes seem to be essentials. Several inverted repeats elements were described in the promoter regions and suggested to be involved in the oxygen-dependent transcriptional regulation of the *nrdIEF* and *nrdDG* genes [127] and further work needs to corroborate such hypothesis.

In *Streptomyces* with class Ia and II RNRs, the *nrdAB* genes are co-transcribed with *nrdS*, which encodes an AraC-like regulatory protein. A knockout mutant of the *nrdS* gene had no detectable effect on the transcription of either set of RNR genes [128]. The class II *nrdJ* gene forms an operon with *nrdR*, a putative transcriptional regulator encoding a protein with an ATP-cone domain similar to the one present in the allosteric overall activity site of class Ia and III RNRs [128] (see below). Quantitative analyses of the transcription of the two RNR operons in *Streptomyces* revealed that the *nrdJ* transcription was constant over the entire course of exponential growth, whereas the *nrdAB* transcription was much lower than that of *nrdJ* already at the early stages and was further markedly decreased in the later stages. It was proposed that class Ia RNR operates primarily in the early stages of growth following spore germination, whereas the class II RNR acts primarily during vegetative growth [129] and also to provide the dNTPs necessary for DNA repair during oxygen limitation [128]. Knockout mutants of either RNR class had no discernible effect on growth individually, but elimination of both RNR systems was lethal for *Streptomyces*. Interestingly, a mutation in the *nrdR* gene causes a dramatic increase in the transcription of *nrdJ* and to less extent of *nrdAB*, indicating that the NrdR regulator is involved in the global regulation of the two RNRs present in *Streptomyces* (see below).

Riboswitches in RNR Operons

A computational study by Gelfand and co-workers identified a regulatory system common to many bacterial RNR classes [130]. They found a highly conserved RNA structure, called B_{12}-riboswitch, which is widely distributed in the 5'-untranslated region (5'-UTR) of vitamin B_{12}-related transcripts including several coding for RNRs. A riboswitch is an intricately folded RNA domain that serves as a ligand-responsive control element that modulates the transcription of certain mRNAs in response to changing concentrations of the metabolite (ligand). Vitamin B_{12} is the metabolite for the B_{12}-riboswitch; in the absence of B_{12} transcription of the gene proceeds as normal, but when B_{12} is bound to the 5'-UTR region of the nascent transcript a conformational change inhibits further downstream transcription (Figure 9). For general information on riboswitches we refer to some recent reviews [131-133].

Surprisingly, B_{12}-riboswitches were not only found in the 5'-UTR regions of *nrdJ* genes (coding for the B_{12}-dependent class II RNR) but also in the upstream regions of other RNR operons, and in some organisms only in the other RNR operons and not in the B_{12}-dependent class II operon. This is the case in *S. coelicolor, M. flagellatus* and *Bacillus halodurans* where the B_{12}-riboswitch sequence is found in the *nrdAB* operon, but not in the *nrdJ*, in *Brucella melitensis, Agrobacterium tumefaciens* and *Bacillus stearothermophilus* where the B_{12}-element is found in the *nrdHIEF* operon and not in the *nrdJ* operon, in

Desulfitobacterium hafniense and *Porphyromonas gingivalis* where the B_{12}-element is found in the *nrdDG* operon and not in the *nrdJ* operon, and in *Bacteroides fragilis* where the B_{12}-element is found in the *nrdAB* operon and in the *nrdDG* operon but not in the *nrdJ* operon. On the other hand, in *Chlorobium tepidum* that only codes for an *nrdJ* gene, its 5'-UTR region contains a B_{12}-element. These observations indicate that B_{12} availability plausibly determines which RNR class will be expressed. High concentrations of vitamin B_{12} allow optimal enzyme activity of the B_{12}-dependent class II RNR and signal repression of the other RNR classes via the B_{12}-riboswitch.

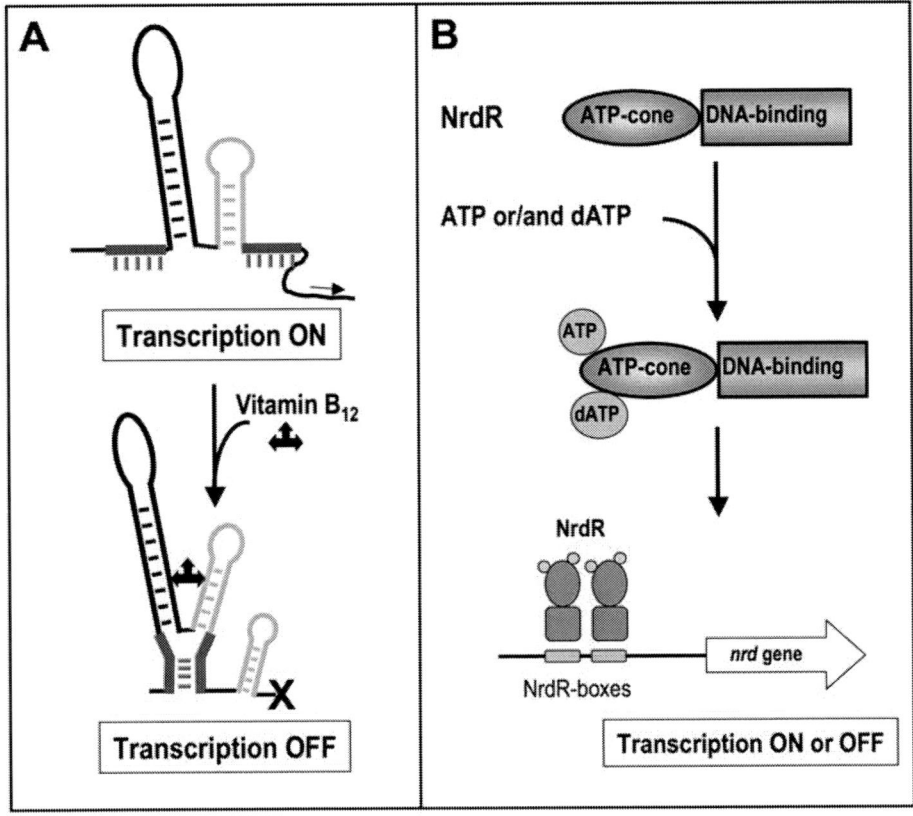

Figure 9. Regulation of *nrd* expression by riboswitches and NrdR. In A) is shown an RNA molecule with a B_{12}-riboswitch and its modulated inhibition of an *nrd* gene in the presence of the vitamin B_{12} ligand. In B) is shown a possible mechanism for transcriptional regulation of an *nrd* gene regulated by NrdR. After binding of ATP or dATP the NrdR protein can modulate the expression of the *nrd* genes by binding to the NrdR box.

The B_{12}-riboswitch regulation of RNR genes has so far only been studied in *S. coelicolor* [134]. The 5'-UTR of the *nrdABS* region contains a 123-nt B_{12}-riboswitch. Already low concentrations of B_{12} inhibit the transcription of the *nrdABS* genes, and in an *nrdJ*-mutated background bacterial growth is completely inhibited in presence of B_{12}.

This is a smart way to balance the expression of the two RNR operons in *Streptomyces*, with vitamin B_{12} controlling the expression of class Ia RNR genes via the riboswitch and serving as a cofactor for the B_{12}-dependant class II RNR.

NrdR - A Global Bacterial Regulator

Our description of different mechanisms that regulate the expression of the three RNR classes in bacteria has so far had very little information on how the expression might be coordinated within a single microorganism with several RNR classes. A recent computational study by Gelfand and coworkers highlights components of a regulatory mechanism [135] that may actually serve such a global coordination. Using comparative genomics they analyzed the promoter regions of all know ribonucleotide reductase genes in bacterial genomes and localized a highly conserved palindromic signal with the consensus sequence acaCwAtATaTwGtg [135]. It was named NrdR box, and this sequence was found in 243 Eubacterial RNR operons, but was lacking in some bacterial genomes and not present in Archaea or in eukaryotes. The NrdR-box can occur as a single box, but more often as tandem boxes. Conservation of the signal along all bacterial domains suggested a universal regulatory mechanism.

Using an extended phylogenetic search Gelfand and coworkers identified only one putative open reading frame that was present in Eubacteria but absent in Archaea and eukaryotes [135]. The potential transcriptional regulator, named NrdR, contains approximately 150 amino acid residues including a DNA-binding Zn-ribbon domain and an ATP-cone domain. It is the same regulatory protein as described in the *Streptomyces* studies carried out by Borovok et al. [128]. The ATP-cone domain is highly similar to the one found in the allosteric overall activity site in class Ia and III RNRs and is in NrdR plausibly involved in sensing the ATP/dATP ratio to differentially regulate the expression of RNR operons. The NrdR protein was represented in all Eubacteria except in the genomes of *Bacteroidetes/Chlorobi* and ε-proteobacteria. In other proteobacteria the NrdR protein is often clustered with genes for riboflavin biosynthesis *(rib)*, glycine metabolism *(glyA)* and transcriptional antitermination *(nusB)*. In other Eubacteria

NrdR clusters together with RNR genes or with genes involved in DNA replication (*dnaA, dnaI, polA*).

The first experimental evidence demonstrating that the NrdR protein is a transcriptional factor regulating RNR expression comes from the study of the *S. coelicolor* NrdR protein [136]. Purified NrdR protein is a tetramer that binds 0.7-0.8 zinc ions per polypeptide. A deletion of the N-terminal Zn ribbon domain abolished zinc binding. The NrdR protein can also bind both ATP and dATP at the ATP-cone domain [136]. NrdR represses the transcription of the two sets of RNR genes (Ia and II) in *S. coelicolor* by binding to the NrdR-boxes upstream of each promoter region. Interestingly a truncated NrdR protein without the ATP-cone is not able to bind to the promoter regions of the *nrd* operons suggesting that this domain can modulate the capacity of the NrdR protein to bind to DNA [136].

A recent collaborative study has shown that the *E. coli* NrdR protein can bind to the NrdR-boxes found in the promoter regions of the three *E. coli* RNR classes (Ia, Ib, III) [137]. In addition a knockout mutant in the *nrdR* gene caused a dramatic increase in transcription of the *nrdHIEF* operon, suggesting a major role of the NrdR protein in the transcriptional regulation of class Ib RNR. Obviously, more experiments are needed to elucidate to what extent the NrdR protein is a global regulator for all RNR genes in bacteria. It is necessary to identify under which environmental conditions NrdR can modulate the transcription of one set of RNR genes in a single bacterium and not the other set(s). In addition, the outstanding question which nucleotide that is bound to the ATP-cone at different environmental cues will open new avenues to understand how one RNR and not the other is expressed to match each specific environment that the organism encounters.

RNR Gene Regulation in Saccharomyces Cerevisiae

Early studies of RNR in budding yeast identified three different genes, two encoding the large subunits (*RNR1* and *RNR3*) [51] and one encoding the small subunit (*RNR2*) [9, 10]. Surprisingly, pure Rnr1 plus Rnr2 recombinant proteins had no enzyme activity and no tyrosyl radical was detected. Homology comparison with the *RNR2* gene discovered a second copy for a small subunit, named *RNR4* [138, 139]. Despite high homology to *RNR2*, *RNR4* lacked several conserved residues involved in formation of the essential iron center. It was therefore quite unexpected to find that Rnr4 in combination with Rnr1 and Rnr2 was essential for RNR activity. Rnr4 interacts strongly with Rnr2 and the active form of the yeast small subunit is the Rnr2/Rnr4 heterodimer. But the exact role of Rnr4 in the RNR holoenzyme is still debated [52, 53] with suggestions that Rnr4 is needed for Rnr2 folding or to allow iron binding and tyrosyl radical formation in Rnr2.

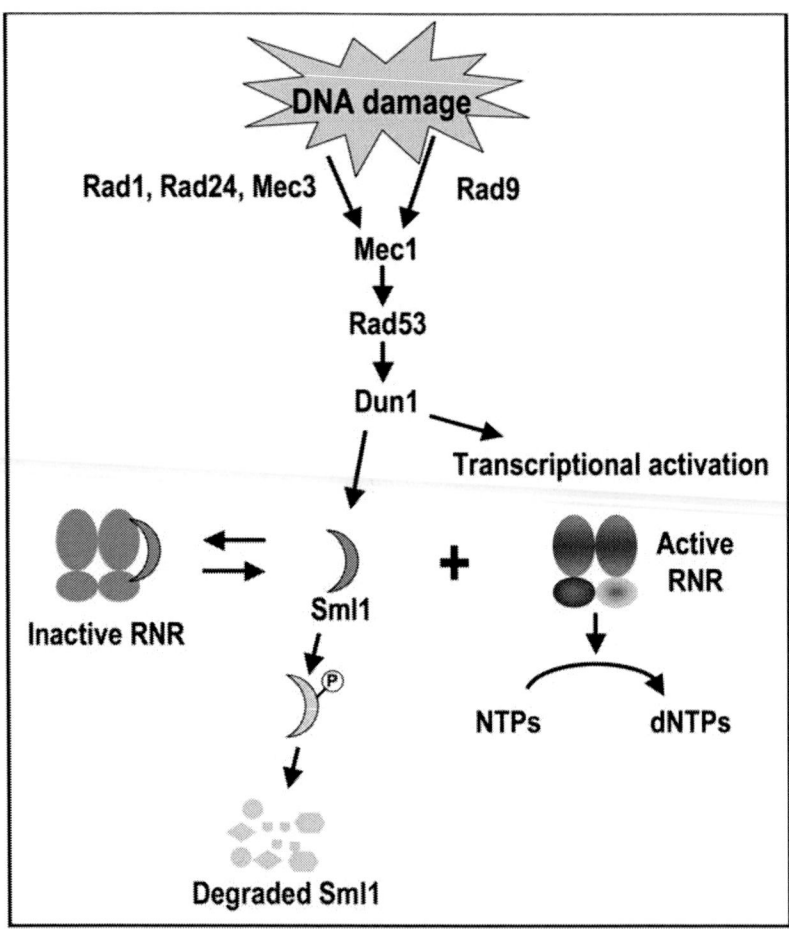

Figure 10. Proposed model for RNR regulation *S. cerevisiae*. It is important to note that the phosphorylation and further degradation of Sml1 drives the equilibrium of the inactive and active RNR complex. The figure is an adaptation from the mechanisms proposed by Zhao et al. [141].

The four RNR genes in *S. cerevisiae* are not coordinately regulated. Instead the *RNR2* and *RNR4* genes are constitutively expressed during the cell cycle, the *RNR1* gene is S-phase-specifically expressed, and the *RNR3* gene is not expressed under normal conditions but is highly induced upon DNA damage. The main target for RNR regulation is *RNR1* which is both transcriptionally regulated and posttranslationally inhibited in the G1-phase by interacting with Sml1 [140]. The levels of Sml1 protein are opposite to the levels of Rnr1, high at G1 and low in S phase, thereby controlling the holoenzyme availability [141].

In response to DNA damage or low nucleotide levels, the transcriptional rates of the RNR genes are rapidly elevated through the activation of the DNA replication stress response pathway, including the *MEC1-RAD53-DUN1* kinase cascade. Sml1 is phosphorylated by Dun1 and is degraded in S-phase and also after DNA damage in a checkpoint dependent manner to relieve RNR inhibition [141]. Recent evidence suggests that an additional regulatory mechanism controls RNR activity by differential localization of its subunits. Wmt1 is a protein that acts as a nuclear anchor to maintain the Rnr2/Rnr4 heterodimer in the nucleolus when cells are outside the S-phase, while the Rnr1 is located in the cytoplasm. At low dNTP levels or after DNA damage the Wmt1:Rnr2/Rnr4 association is disrupted and the Rnr2/Rnr4 heterodimer becomes cytoplasmic, where it can interact with Rnr1 to form an active holoenzyme complex [142]. A summary of the RNR regulation in *S. cerevisiae* is shown in figure 10.

Chapter V

Mutational Studies in RNRs

The essential role of RNR in cell proliferation was first established in *E. coli* in the early 1970ies, when a large mutational screen of bacteria with temperature sensitive DNA synthesis resulted in isolation of the *dnaF* mutation [143]. The mutation was subsequently shown to affect the enzyme activity of the large subunit of class Ia RNR and the gene was named *nrdA* [144]. DNA sequencing studies in the authors' laboratory could never confirm any mismatch mutation in the coding region of the *nrdA* gene, and it is plausible that the mutation affects expression of the gene (Regnström, K. and Sjöberg, B.-M., unpublished). Later two independent conditional mutations in *nrdB* were also isolated [145-148], and they were in the 1990ies shown to encode a L304F and P348L mismatch mutation respectively in *E. coli nrdB* [149]. As has been discussed above, only conditional mutants were obtained since the chromosomal *nrdEF* locus could not substitute for defects in the *nrdAB* locus. Another early non-directed mutation in mammalian class Ia RNR was isolated as a dATP resistant phenotype and shown to be a D57N mismatch in the allosteric overall activity domain in mouse NrdA [150]. Since then site-directed mutational studies of RNRs have been concentrated on understanding the reaction mechanism, the allosteric regulation, and recently also protein-protein interactions (Table 3). With the appearance of facile methods for gene knockout even in multicellular eukaryotes, several recent studies have aimed to elucidate whether some RNR genes are optional in organisms encoding more than one set of RNR genes (see above). In addition recent SNP (single nucleotide polymorphism) analyses have sought to find out whether RNR mutations may be involved in certain diseases, in particular in different forms of cancer [151-156].

Table 3. Site-directed mutations in the RNR proteins

Organism	Gene	Mutations	Effects	Ref
Chlamydia trachomatis	nrdA	Δ1-248	loss of dATP inhibition, no general effect on enzyme activity	[62]
	nrdB	F127Y	no tyrosyl radical, despite tyrosine introduced at consensus position	[236]
		Y129F	loss of enzyme activity	
		F127Y/Y129F	no enzyme activity	
		H59A, H59N, H59Q, H88A	impaired dATP inhibition	
		C225A, C225S, C462A, C462S, C439A, C439S	no enzyme activity, affect active site cysteine pair	[29, 30]
		C754A, C754S, C759A, C759S, C754S/C759S	no enzyme activity, affect site of transient thiyl radical	
Escherichia coli	nrdA	N437A, N437D, N437Q, N437S, E441A, E441D, E441Q	no enzyme activity, defective reaction mechanism	[237, 238]
		Y730F, Y731F	no enzyme activity, defective radical transfer	[239]
		N238A, V245K, H284F, H284V, T287N	defective (impaired) NrdA dimer interaction	[240]
		W48A, W48F, W48Y, D237E, D237N, Y356A, Y356F, Y356W D84A, D84E, D84H, E115A, H118A, E204A, E204H, E238A, H241A	no enzyme activity, affect radical transfer pathway	[166-171]
	nrdB		defective iron bindning, no (decreased) enzyme activity	[163, 164]
		Y122F, Y122H	no tyrosyl radical, no enzyme activity	[32, 160]
		F208Y, F212Y, F212W, I234N	defective radical stability, affect environment of tyrosyl radical	[165]
		Δ30C	no enzyme activity, no interaction with NrdA	[241]

Organism	Gene	Mutations	Effects	Ref
Escherichia coli	nrdD	G681A	no glycyl radical, no enzyme activity	[158]
	nrdG	C26A, C30A, C33A	impaired metal binding, no enzyme activity	[49]
		C119S, C419S	no enzyme activity, affect active site cysteine pair	
Lactobacillus leichmannii	nrdJ	C408S	no enzyme activity, affects proposed transient thiyl radical position	[42]
		C731S, C736S, C731S/C736S	no thioredoxin interaction	
Pseudomonas aeruginosa	nrdA	Δ1-147	loss of dATP inhibition, no general effect on enzyme activity	[27]
		Δ1-182	no enzyme activity	
Ralstonia eutropha	nrdD	G650A	no enzyme activity	[66]
		N78A, N78C, N78D, N311A, N311C, N78A/N311A	defective enzyme activity	[242]
		C79S, C290S	no enzyme activity, affect active site cysteines	
		C260S, C453S, C579S	no (decreased) effect on enzyme activity	[45]
Bacteriophage T4	nrdD	C543S, C546S, C561S, C564S	no enzyme activity, no glycyl radical, affect Zn-finger motif	
		G580A	no enzyme activity, affects site of glycyl radical	[159]
		F194A	defective allosteric regulation, decreased enzyme activity	[243]
		M288A, M288C	no enzyme activity	
		R291A, R291E, R291Q	loss of dATP inhibition, no general effect on enzyme activity	
Mus musculus	nrdA	D57N	no phosphorylation, no general effect on enzyme activity	[244]
	nrdB1	S20A, S20D		
		K30A/E31A	no ubiquitin-depenent degradation in late mitosis, mutation of KEN box	[245]

Table 3. Continued

Organism	Gene	Mutations	Effects	Ref
Mus musculus	mrdB1	W48Y, W48F, D266A, Y370F, Y370W	defective radical transfer pathway	[172-174]
		Y177F, Y177C, Y177W, Y177F/I263C	no tyrosyl radical	[161, 162]
Saccharomyces cerevisiae	nrdA1	C428A	no enzyme activity, affects proposed transient thiyl radical position	[52]
	nrdB2	Δ8C	no interaction with NrdA1	[246]

Table 4. Substrate analogues as inhibitors of RNR

Type of inhibitor	Name of compound	Synonym	Acts on	Comments	Clinical stage	Ref
Substrate analogue	Gemcitabine	Gemzar, 2',2'-diflouro-deoxycytidine, dFdC	NrdA & NrdB		Approved	[180, 188, 189, 197]
	Tezacitabine	(E)-2'-deoxy-2'-(fluoromethylene) cytidine, FMdC	NrdA & NrdB		Phase II	[194, 195]
	DMDC	2'-Deoxy-2'methylidene cytidine	NrdA & NrdB		Phase I	[196, 197]
Effector	Cladribine	2'-Chloro-2'deoxyadenosine, CdA	NrdA		Approved	[247]
	Fludarabine	9-β-D-arabinofuranosyl-2-fluoroadenine 5'-monophosphate, FaraA	NrdA		Approved	[232]

Type of inhibitor	Name of compound	Synonym	Acts on	Comments	Clinical stage	Ref
Effector	Clofarabine	2'-Chloro-2'-fluoro-2'deoxy-9-β-D-arabinofuranosyl-adenine, Clolar, Evoltra, Cl-F-araA	NrdA		Approved	[233, 234]
Radical scavenger	Hydroxyurea	Hydrea, hydroxycarbamide	NrdB		Approved	[248]
	4-hydroxy-anisole	Para-metoxyphenol, mequinol	NrdB		Approved	[249]
	Trimidox	3,4,5-Trihydroxy-benzamidoxime	NrdB		In test in combination treatments	[201]
	Didox	3,4-Dihydroxy-benzohydroxamic acid	NrdB		Phase II	[201]
	Resveratrol	3,5,4'-Trihydroxy-stilbene	NrdB		*In vitro*	[202, 203]
Iron chelator	Deferoxamine	Desferal, DFO	NrdB	Indirect by iron chelation	Approved	[208, 209]
	Exochelin	Desferri-exochelin, D-Exo	NrdB		Cell line, *in vitro*	[213]
	Triapine®	3-Aminopyridine-2-carboxaldehyde thiosemicarbazone, 3-AP	NrdB		Phase II	[214, 250]
Other	Cisplatin	Cis-diamminedichloroplatinum, Platinol	NrdA & NrdB	Main action on DNA	Approved, often used in combination treatments	[225, 226]
	Caracemide	*N*-acetyl-*N,O*-di(methylcarbamoyl)hydroxylamine	NrdA	Too many toxic side effects	Phase II	[228]
	Cytarabine	Arabinosylcytosine, AraC	NrdA?	Main action on DNA	Approved	[229, 251]
	Nelarabine, Arron	Prodrug converted to arabinosylguanosine, AraG in the cell	NrdA?	Main action on DNA	Approved	[231]

The first site-directed mutation to be engineered in an RNR gene, aimed at identifying the position of the stable tyrosyl radical in the NrdB protein of class Ia RNR from *E. coli* [32]. Several years prior to the elucidation of the three-dimensional structure for the NrdB protein, these studies were based on amino acid sequence alignments of NrdB proteins from *E. coli*, a marine mollusk, and two eukaryotic viruses [157]. The mutation Y122F abolished the tyrosyl radical and the enzyme activity (Table 3).

The three active site cysteines were identified in a similar manner to Cys225, Cys439, and Cys462 in the NrdA protein of *E. coli* class Ia RNR [29, 30] and a few years later to Cys119, Cys408 and Cys419 in the NrdJ protein of class II RNR from *L. leichmannii* [42]. In class III RNRs the position of the stable glycyl radical was assigned to Gly681 in *E. coli* NrdD [158] and Gly580 in bacteriophage T4 NrdD [159], and subsequently the two active site cysteines were assigned to Cys79 and Cys290 in the T4 NrdD protein [45].

The stable tyrosyl radical has subsequently been engineered to other side chains in *E. coli* NrdB [160] and in mouse NrdB [161, 162]. The primary aim has been to introduce novel potential radical harboring side chains that might help in understanding how the tyrosyl radical is formed and details of the enzymatic reaction mechanism (Table 3). In *E. coli* NrdB these studies have also been complemented with extensive engineering studies of the side chains that ligate the diiron site [163, 164], the hydrophobic side chains in the surroundings of Tyr122 [165], and in side chains presumed to participate in electron transfer processes during oxidation of the diiron site and the concomitant generation of the tyrosyl radical [166-171]. These studies have collectively contributed to the consensus view that the radical site in NrdB can be either a tyrosyl radical stabilized by a diferric-oxo site or a high valent diiron (Fe^{III}-Fe^{IV}) site [62, 70], and that oxidation of the diiron site requires oxygen and a dedicated electron transfer pathway (cf. Chapters 5-8).

The exceptionally long distance between the tyrosyl radical in NrdB and the active site cysteine in NrdA has triggered a series of engineered mutational studies in *E. coli* and mouse class I RNRs [166-174]. In addition, recent elegant studies with specifically introduced unnatural amino acid side chains have taken these studies one step further and demonstrated that Tyr356 at the surface of NrdB participates as a redox-active side chain in the radical transfer process to NrdA [175].

The successful elucidation of several steps in the reaction mechanism has relied upon the advantageous combination (primarily in *E. coli* class Ia RNR) of engineered mutations (Table 3) and substrate analogues [176-180], and has formed a basal platform for initiating cell culture experiments and clinical trials with several of the antiproliferative drugs described below.

Chapter VI

RNR as an Antiproliferative Target for Disease Control

To use RNR as a target for drugs in cancer, virus and parasite combat is an attractive approach. However it is not without complications since a drug that is not specific enough will also interfere with other systems giving toxic effects to the patient. If, on the other hand, the drug is highly specific it is often found that the inhibition of RNR is still not sufficient to stop proliferation. Administration of the drug is also a bottleneck, oftentimes the drug needs to be modified in the cell to perform its inhibitory action. Hence the promising results obtained *in vitro* are frequently not obtained *in vivo*. Further complications arise from the fact that different cell lines respond very differently to the same drug and that resistance develops quite frequently. These factors will be touched upon in the below description of different chemicals developed to inhibit RNR activity. Despite all these difficulties large efforts have been put into curing cancer, virus infections and pathogen attacks by inhibiting RNR, as is reflected in the more than 460 patents targeting RNR that have been registered in the last 30 years [181]. Some of the compounds that are approved, in clinical trials, or are therapeutically promising candidates are listed in table 4.

Since RNR is a complex enzyme under strict regulation different parts of the enzyme can be targeted in drug-design:

- substrate analogues will usually inhibit irreversibly,
- radical scavengers destroy the tyrosyl radical,
- iron chelators remove the iron needed to form the diiron-radical site,

- peptidomimetics inhibit the protein-protein interaction between NrdA and NrdB,
- antisense techniques, RNA interference and gene therapy manipulate the expression of RNR,
- effector analogues and other potential inhibitors usually decrease enzyme activity.

Substrate Analogues

Substrate analogues that inhibit RNR are primarily modified in the 2' position of the ribose (Figure 11, table 4), and the most common analogues are the halogenated substrates but also azidoNDP has been much studied [176-180]. *In vitro* studies with the class Ia model proteins from *E. coli* and in some cases also class II RNR [182, 183] have shown that the common denominator is a disturbance of the delicate radical reaction that goes astray and prevents regeneration of the essential cysteinyl radical and thereby in the class I enzyme also results in tyrosyl radical decay (see Figure 1).

A difference between *in vitro* and *in vivo* studies is that the substrate analogue can be added in the proper phosphorylated form (di- or tri-) *in vitro*, whereas it is the nucleoside analogue that is administered to the cells or added to the medium *in vivo* and has to be phosphorylated by kinases before it can act as an inhibitor. Hence both di- and triphosphorylated forms of the analogue will be present in the cell. In mammalian cells the diphosphate form can inhibit RNR whereas the triphosphate form may be incorporated into DNA by polymerase and stop chain elongation.

Cells are prone to circumvent cells death by evolving defense mechanisms that frequently gives rise to resistance against a drug. Different cell lines respond very differently to a particular drug. Resistance to nucleoside analogues can occur at many levels; nucleoside transporters that provide the uptake and transport of the analogue into the cell can become less efficient, nucleoside and nucleotide phosphorylation can become less efficient, deaminase and phosphatase processes reduce the levels of the analogue, and overexpression of RNR decreases the relative intracellular levels of the analogue [184-187].

Gemcitabine (2',2'-difluorodeoxycytidine, dFdC) is a substrate analogue with two fluorines attached to the 2' of the ribose ring (Figure 11). This analogue that has been approved for cancer therapy will be described in some detail since it can stand as a prototype for this type of drugs. Furthermore it has been subjected to a number of biochemical and biophysical studies *in vitro* as well as to crystallization [180, 183, 188].

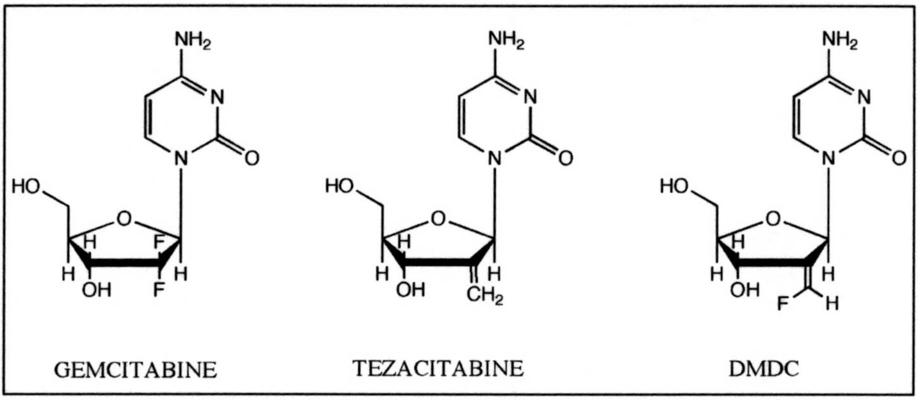

Figure 11. Substrate analogues inhibiting RNR.

With *E. coli* class Ia RNR it was shown that the NrdA protein is inactivated, the tyrosyl radical decays, two fluorides and one cytosine are released per inactivated RNR, and that the kinetics for fluoride release and tyrosyl radical decay are coupled. The NrdA is thought to be inactivated by a covalent modification but the site has not been identified even if it is thought to be within the active site [180]. The fate of the radical has not been elucidated, but a furanone adduct usually formed with other 2'-substituted inhibitors is not formed in this case [189]. Instead a non-stoichiometric amount of a substrate-derived radical (15-25% of the decayed radical) was observed during the inactivation [180]. Gemcitabine has been crystallized bound to NrdA1 (Rnr1) of *S. cerevisiae* class Ia RNR where the structure was compared to bound CDP [188]. Obviously the two fluorines change the binding pattern of the side chains in the active site but there are also two water molecules coordinated to residues in the active site that presumably change the reaction of the 3'-hydroxyl group of the ribose (Figure 12). A theoretical study on the mechanism for gemcitabine inactivation of *E. coli* RNR suggested that the reaction followed the natural reaction pathway up to formation of the disulphide bridge at the active site but deviates after that [190]. The study indicates that future substrate analogues can profit from having two good leaving groups in the 2' position and that the substituent in the α-face of the inhibitor must leave in protonated form which is the case for chlorine and fluorine substituents. It would be interesting to see if the outcome of the calculations would be different if based on the eukaryotic structure [188].

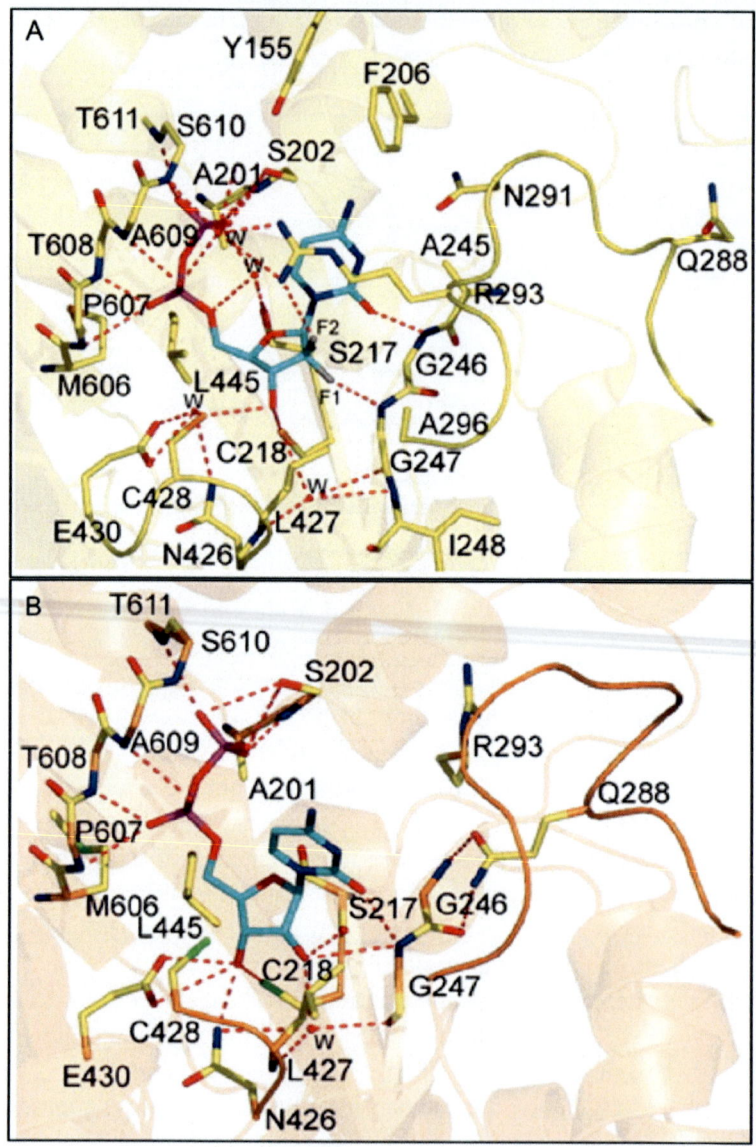

Figure 12. Gemcitabine (A) and CDP (B) binding to *S. cerevisiae* RNR1 [188]. The active site residues interacting with the inhibitor/substrate differ in the two structures. In particular residues Glu430 and Asn426 that have been shown to be important in the reaction have different coordinations and are not able to form hydrogen bonds to the 3'-OH in the gemcitabine structure. Instead a water molecule intersecting Glu430 and 3'-OH may be important for the gemcitabine reaction, and there is a lack of the water molecule seen in the CDP structure that has been implicated in the normal substrate reaction.

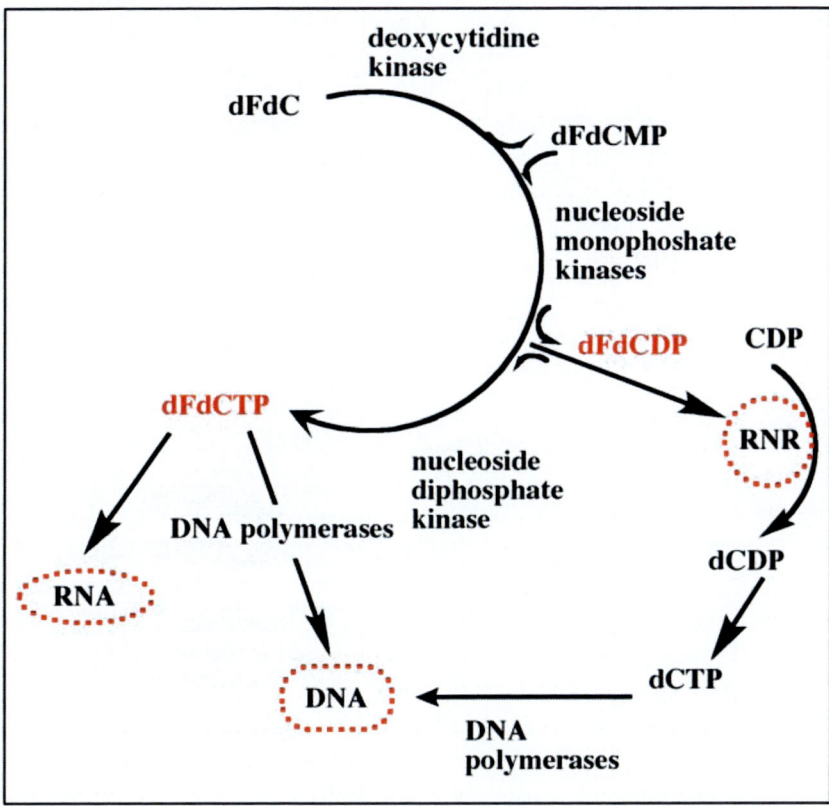

Figure 13. Activation of the RNR inhibitor gemcitabine and its inhibitory roles in the cell. Gemcitabine, dFdC, is phosphorylated by three kinases to dFdCMP, dFdCDP and dFdCTP. In red are shown the metabolites that have inhibitory effects and the targets of their action are surrounded by hatched red borders.

Gemcitabine inhibits both NrdA (M1) and NrdB (M2) activities in the human system (Stubbe, personal communication in [188]). In therapy the drug is administered as the nucleoside dFdC, which is metabolized to the diphosphate form by two different kinases (Figure 13), and further phosphorylated to the triphosphate form. In one study the dFdCDP level was only 2-3% of the dFdCTP level in treated cells [191] and this can be the reason why gemcitabine has cytotoxic effects in non-dividing cells which has been attributed to the effect of dFdCTP on RNA metabolism. Despite the efficiency of gemcitabine to inhibit RNR it is not a complete success in cancer treatment and ongoing clinical trials test it in combination with other drugs. The observed resistance to gemcitabine could occur on different levels as discussed above and different cell lines show different behaviors [184, 187, 192].

For instance a human mammary adenocarcinoma cell line with 500-fold increased resistance to gemcitabine had an increased expression of *nrdA* but not of *nrdB*. The cells also had an increased resistance to arabinosylcytosine (araC, see below) and HU [192]. Increased levels of NrdA have also been observed in mouse leukemia, non-small-cell lung cancer, and oropharyngineal carcinoma cells. One approach to overcome resistance could explore the finding of Jordheim et al [193], who showed that gemcitabine resistance due to deoxycytidine deficiency could be reverted by transfection with *D. melanogaster* multisubstrate deoxynucleoside kinase in a human uterine sarcoma gemcitabine resistant cell line.

Tezacitabine, (E)-2'-deoxy-2'-(fluoromethylene) cytidine (FMdC), is a substrate analogue that has reached phase II trials for solid cancers (Table 4). It is an irreversible inhibitor of RNR, inhibits also DNA synthesis after incorporation into DNA but has the advantage of being less susceptible to deamination [194]. Its reaction with *E. coli* RNR has been thoroughly investigated and it reacts similarly to gemcitabine. It inactivates both NrdA and NrdB, a substrate radical is observed and alkylation of an active site residue is considered to be a likely result of the reaction [195].

DMDC, 2'-deoxy-2'methylidenecytidine is a drug that has passed phase I trials. The advantage of this analogue over gemcitabine and araC is that the cytidine deaminase activity, normally a drawback in therapy, here lowers the concentration of 2'-deoxycytidine in the cell reducing the concentration of the competitor to DMDC to be phosphorylated by cytidine kinase in the cell. Thereby the concentration of the active form of the drug is increased in the cell. Since many cancers that are difficult to treat have high activity of their cytidine deaminase this is a promising approach [196, 197].

Radical Scavengers Attacking the Tyrosyl Radical

The first antiproliferative drug to be used against RNR was *Hydroxyurea* (HU) and this radical scavenger that reduces the tyrosyl radical is still a standard drug in treatment of many cancers (Figure 14). However HU is rather inefficient and high concentrations need to be administered with undesired side effects for the patient. One such is bone marrow toxicity. Due to the toxicity HU is often combined with other drugs, and frequently used in HIV and AIDS treatment in combination with AZT or 2',3'-dideoxyinosine (ddI). HU is also tested with good results in sickle cell anemia [198]. Another radical scavenging chemical used in therapy is *4- hydroxyanisole*, which is used against malignant melanoma. That it can be useful to apply relatively simple radical scavenger on pathogenic bacteria is reflected by our recent discovery that *N-methylhydroxylamine* turned out to be several hundredfold more effective against *Bacillus anthracis* NrdF as compared to NrdB from the mammalian host [199]. Most likely we will find use for the difference in reactivity towards NrdBs from different species in future drug design.

Trimidox (3,4,5-trihydroxybenzamidoxime) and *didox* (3,4-dihydroxybenzohydroxamic acid) are both good radical scavengers and can also form iron complexes. Trimidox and didox are 100 and 20 times more efficient in radical scavenging *in vitro* than HU [200].

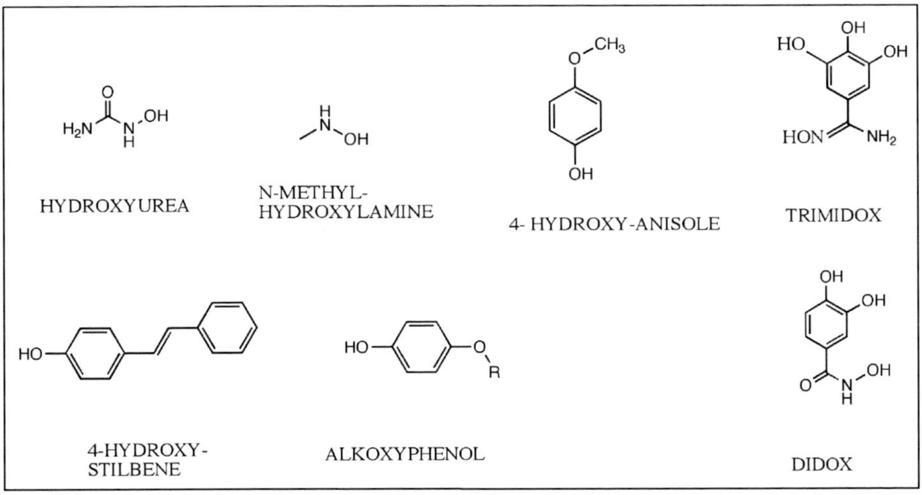

Figure 14. Radical scavengers inhibiting RNR.

Both compounds have been tested in combination therapy with ddI, and are efficient in suppressing murine retrovirus-induced immunodeficiency. A benefit with these drugs is that they do not cause bone marrow toxicity and that they lower both the purine and pyrimidine dNTP pools whereas HU only reduces the purine dNTP pools [201].

Resveratrol, 3,5,4'-trihydroxystilbene, is a compound from grapes thought to protect the plant from stress and pathogen attacks. It inhibits processes involved in tumor initiation, promotion and progression [202] and since it was also found to have low toxicity it was suggested to have potential in chemotherapy. Fontecave et al. have shown that resveratrol is a potent inhibitor of RNR [203]. Resveratrol was ca 25-fold more efficient than HU in inhibiting DNA synthesis in mouse lymphoblast leukemia cells, which were overexpressing NrdB 15-20 fold, and it was also a more efficient inhibitor than the very potent drug p-propoxyphenol (see below). That RNR was the target was shown on pure mouse NrdB where the radical decay with resveratrol was much more potent than with HU.

The importance of stereospecificity was emphasized in a study where tyrosyl radical decay in mouse NrdB was compared between the related drugs resveratrol, *4- hydroxy-trans-stilbene* (4HTS) and 3-hydroxy-trans-stilbene (3HTS). 4HTS was slightly more efficient in tyrosyl radical reduction compared to resveratrol whereas 3HTS did not scavenge the radical at all [204].

Alkoxyphenols are inhibitors that work as radical scavengers of NrdB; p-propoxyphenol belongs to this group. Their mode of operandi differs between species. When they react with mammalian and herpes simplex NrdB the iron center is lost together with the tyrosyl radical, whereas they only reduce the radical in NrdBs with a more closed diiron/radical site as for instance in *E. coli* NrdB. The kinetics of a series of p-alkoxyphenols showed that the radical decay rate increased with the length of the alkyl chain for the mammalian enzyme, whereas there was no simple correlation for herpes simplex virus (HSV) RNR [205]. It is however wise not to be prejudiced as to how efficient a radical scavenger will be with NrdB from different species as for instance observed with *B. anthracis* RNR described above.

Nitric oxide (NO) produced by nitric oxide synthase is an endogenous signaling molecule in cells. NO has also been found to function in macrophage defense mechanisms against parasites infections. Really small molecules, like NO, that can diffuse into RNR can reduce the tyrosyl radical and possibly substrate radicals as well and react with cysteines in the active site. Thionitrates are thought to produce nitric oxide *in situ* [206]. It has been shown that reduction of the tyrosyl radical with nitric oxide is reversible in mammalian systems but the dNTP pools are still depleted in tumor cells upon treatment with NO [207].

Iron Chelators

Iron chelators are attractive concepts in therapies against cancer as well as infections since iron is essential for all cells. Depletion of the iron pool can also be a strategy when a virus or a parasite enters replicative state. The depletion will affect RNR since mammalian NrdB needs iron and oxygen for generation of the diiron/radical site. Pure iron chelators rarely act directly on RNR but some chelators can also reduce the tyrosyl radical. Some iron chelators that have been shown to affect RNR activity are in use or under development (Figure 15, table 4).

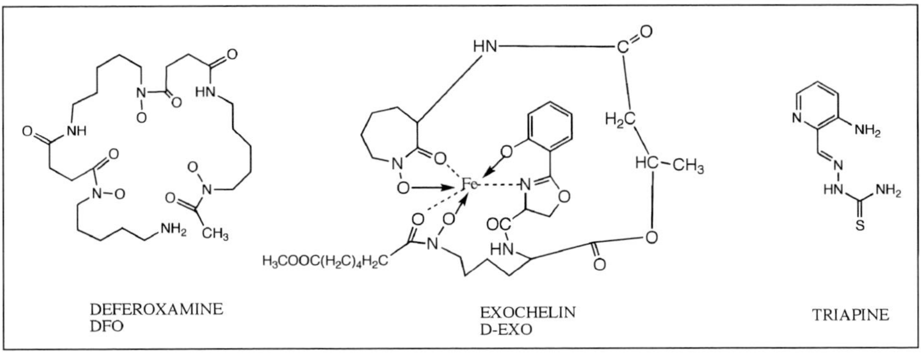

Figure 15. Iron chelators inhibiting RNR.

Deferoxamine is a potent iron chelator used in therapy. It is redox-inactive, the trapped iron is bound in the Fe^{3+} form, and it has been shown not to reduce the tyrosyl radical of mammalian NrdB [208, 209]. A disadvantage of deferoxamine is its short half-life, in blood only 5-10 minutes.

Some efforts to interfere with the function of RNR in an invading parasite have been successful but with certain limits. Inhibiting RNR could theoretically treat *Trypanosoma brucei* infections causing sleeping sickness. *T. brucei* NrdA is present throughout the life cycle and NrdB is expressed in the multiplying state when it is present in blood and tissue fluids and not in the cell cycled arrested phase. NrdB is also expressed in the dividing procyclic *T. brucei* found in the midgut of the Tsetse fly [210]. When the parasite goes into the multiplying phase it needs iron to activate NrdB, aconitase and alternative oxidase. Deferoxamine is 10 times more efficient on *T. brucei* than on mammalian cells, but it does not affect the RNR activity of purified *T. brucei* enzyme [211]. Hence it seems that the drug has to be administered before *nrdB* is expressed to be effective.

The search for easily administrated iron chelators has led to the lipophilic chelators desferri-exochelins. *Exochelins* are excreted from *M. tuberculosis* and have an exceptional binding affinity for ferric iron [212] and according to experiments in cell lysates exochelin is not likely to reduce the tyrosyl radical. Cells were arrested in the G0/G1 and S phases [213].

Thiosemicarbazones inhibit RNR very efficiently by inactivating NrdB but their use as drugs has been limited due to toxic side effects. *Triapine*, 3-aminopyridine-2-carboxaldehyde thiosemicarbazone, is a low toxicity relative to the thiosemicarbazones and a promising drug according to phase II trials in 2006. Triapine inhibits mammalian NrdB and recent studies show that the RNR activity was inhibited with an IC_{50} of ca 0.1 µM, which is almost 10 000 times lower concentrations than with HU [214]. The mode of action has not been investigated in depth biochemically, hence it is not clear if it also can reduce the tyrosyl radical. In phase I tests against advanced leukemia triapine was shown to be well tolerated when it comes to toxicities. Triapine is also investigated together with cytarabine in patients with acute leukemia (see below).

Peptidomimetics Corresponding to the C-Terminus of NrdB

Another approach explores that fact that the NrdB interaction with NrdA is mainly via the C-terminus of NrdB. Addition of a peptide that mimics the C-terminus of the NrdB protein inhibits RNR activity by competing with NrdB to form an active RNR complex. Since the C-terminus is species specific this was thought to be a powerful drug to fight viruses without hurting the host. However, administration of the drug has been an obstacle both when it comes to finding the target and to avoid degradation by proteases.

Peptide inhibition has been most studied to find a cure for HSV infections. HSV RNR is required for full expression of pathogenicity in animal models of primary infection as well as for reactivation of latent infections. Initial promising studies already in the mid 1980ies used a nonapeptide corresponding to the C-terminus of the HSV NrdB protein [215, 216]. The field developed into studies on the importance of peptide length as well as individual side chains for binding to the HSV NrdA protein, and then into substitutions to increase affinity and binding strength. As drug delivery is a problem the peptides have not reached clinical use [217]. An interesting system to overcome this problem is a peptide that was produced in the HSV infected cell as a fusion protein with a heat-labile enterotoxin, from which it was released *in vivo* [218]. A crystal structure of the fusion protein with the HSV nonapeptide shows that the enterotoxin has its native structure now with the flexible C-terminus of RNR. In the cell proteases cleave the peptide that is then interacting with the RNR, and the data suggest that enough peptide survives to inhibit the target molecule.

Peptide competition was considered a possible therapy against *T. brucei*, the causative agent of sleeping sickness. Unfortunately, biochemical characterization of the RNR showed that the C-terminus of the pathogen is very similar to that of the host. The similarity enabled the C-terminus of mouse NrdB to bind to *T. brucei* NrdA with the same affinity as to mouse NrdA. Interestingly however, combinations of *T. brucei* NrdA and host cell NrdB or vice versa were enzymatically inactive [219].

Manipulating the Expression of RNR

Antisense inhibitors of the HSV-2 *nrdA* gene have been used to inhibit growth and latency reactivation of the virus [220]. Antisense is also tested in phase I studies on patients that have not responded to treatments for lymphomas and solid tumors.

Antisense against NrdB is also tested with a 20-mer oligonucleotide directed towards a coding region in the mRNA of human NrdB and has reached phase II studies. It has been shown to decrease mRNA and protein levels of the NrdB protein and to be effective against a number of tumors as well as human melanoma cells in an animal model [221]. Other antisense oligonucleotides have been found to inhibit *nrdB* expression in the treated systems giving hope for more antisense therapeutics against RNR in the future.

A new development to silence expression of the *nrdB* gene in mammals is based on siRNA [222]. A delivery system composed of a cyclodextrin polycation was developed to prevent degradation of the synthetic siRNA. So far the treatment was well tolerated in animal tests.

Gene therapy is also worked on. In this case the vector is a replication defective construct of adenovirus carrying the human R1 gene (*rad5-R1*) [223, 224]. Overexpression of mouse R1 leads to suppression of tumorigenicity, transformation and metastatic properties of tumor cells [223]. The mechanism behind this is not understood but an adenovirus based R1-gene therapy against colon adenocarcinoma is a possible treatment in the future.

Other Potential Inhibitors and Allosteric Nucleotide Analogues

Figure 16 shows a few additional drugs that have been clinically tested as RNR inhibitors or are used in combination therapies with drugs against RNR.

Cisplatin is used in many cancer therapies on solid tumors. It is an alkylating agent that acts on DNA but has also been shown to inhibit RNR. Smith and

Douglas demonstrated in 1989 that cisplatin inactivates *E. coli* NrdA irreversibly by alkylating the cysteines in the active site, but they also showed also that the main target is DNA [225]. Later this study was followed by an investigation on mammalian RNR where both NrdA and NrdB were inhibited [226]. It was suggested that a critically positioned cysteine in mammalian NrdB, not present in *E. coli* NrdB, was alkylated in the mammalian protein. However, studies on different cell lines showed that a 100-fold lower concentration of cisplatin was required to stop colony formation than to inhibit RNR activity [226], suggesting that RNR is not an important target for cisplatin *in vivo*.

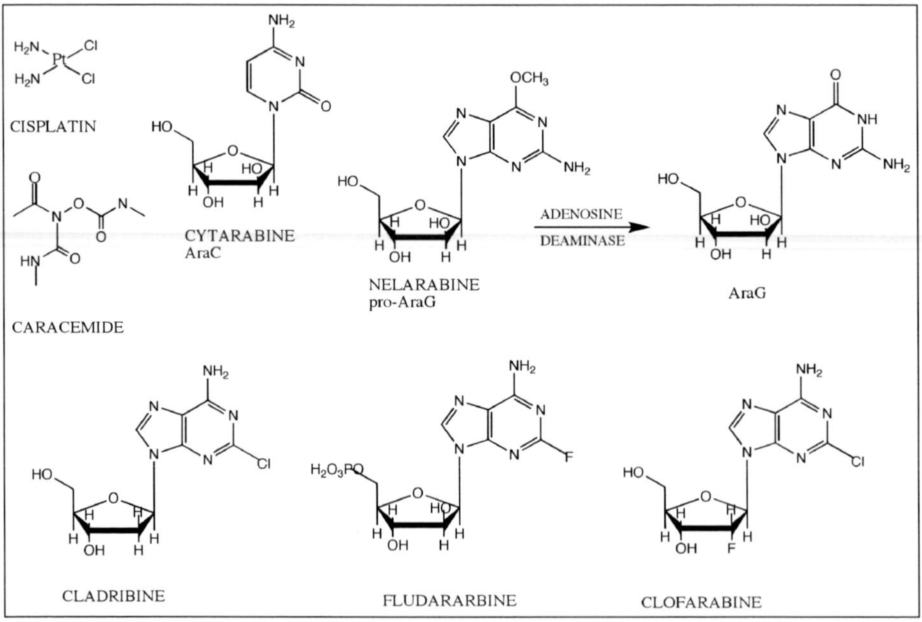

Figure 16. Allosteric nucleotide analogues and other potential inhibitors of RNR.

On the other hand, cisplatin is often combined with other drugs in cancer therapy and synergistic effects are achieved with RNR inhibitors. For instance gemcitabine given to non-small-cell lung cancer and pancreatic cell lines have good synergistic effects with cisplatin [227].

Caracemide was shown to inactivate *E. coli* NrdA irreversibly [228]. It passed into phase II clinical studies but failed due to cytotoxic and neurotoxic side effect and is not used in therapy. Furthermore the instability of caracemide in physiological solutions and blood reduced its use as a drug.

Cytarabine, arabinosylcytosine or araC, is a nucleoside analogue frequently used in cancer therapy to treat acute myeloid leukemia. In the cell araC is metabolized by deoxycytidine kinase (dCK) to the di- and triphosphate forms where the main cytotoxic effect is that araCTP inhibits DNA synthesis after incorporation into DNA. There is some ambiguity in the literature whether araC can inhibit RNR or not, and we have not found any conclusive reports from experiments on pure RNR. However, the araU derivative, 2'-O-allyl-araUDP could inhibit mouse RNR but the corresponding CDP compound had only limited inhibitory effect [229]. AraC has limited applications since it is rapidly deaminated and at high doses araCTP will inhibit dCK by feedback inhibition with neurotoxic effects. Hence it is often combined with known RNR inhibitors in therapy. Cytarabine plus triapine is investigated in patients with acute leukemia in order to enhance the effect of araC by increasing the relative levels of araCTP in the cell. Addition of the NrdB inhibitor will reduce the dNTP pools thereby increasing the incorporation of araCTP into DNA and in that way stop DNA synthesis [230].

Nelarabine, trade name Arran, is a new drug that is a prodrug of arabinosylguanine. The described cytotoxically active form is said to be araGTP in analogy with araCTP but it is also emphasized that additional cytotoxicities are not fully understood [231].

Cladribine, fludarabine and clofarabine act as inhibitors of RNR at the level of allosteric effectors and are all in clinical use [232-234]. Also these nucleoside drugs are phosphorylated in the cell. Cladribine, 2'-chloro-2'deoxyadenosine, inhibits both RNR and DNA polymerase in its triphosphate state in proliferating cells. Fludarabine, 9-β-D-arabinofuranosyl-2-fluoroadenine 5'-monophosphate, inhibits RNR, DNA polymerase, DNA ligase and DNA primase. Clofarabine, 2'-chloro-2'-fluoro-2'deoxy-9-β-D-arabinofuranosyl-adenine, inhibits RNR and DNA polymerase and has a high affinity for dCK. Clofarabine is the most recently approved of the inhibitors in this class under the name of Clolar or Evoltra. As seen in figure 16 clofarabine share the halogenation of the adenine base with cladribine and fludarabine and that confers resistance to deamination. Furthermore it has a fluorine at the 2' position of the sugar that supposedly gives increased resistance to the glycosidic bond.

Chapter VII

Future Antiproliferative Regimes

The central role of RNR in DNA replication and repair will certainly promote its use as an antiproliferative target in the future. With the massive burst of new "high-throughput" and bioinformatics data made publicly available the potential "RNR fingerprint" of a hazardous organism can be predicted and screened during conditions mimicking the infectious or perilous environment. A principal problem with several of the currently used antiproliferative drugs against eukaryotic RNR is that even though they are specific for RNR they cannot differentiate between RNRs in healthy cells and in tumor cells, thereby restricting their clinical use. Future work will have to focus on modifying such drugs to be highly specific against RNR and with reduced side effects. New frontiers where we predict that RNR inhibition will become central are instead parasitic infections in man, and in crops and cattle. One promising recent study in the authors' laboratories identified N-methylhydroxylamine as a potential lead compound for treatment of anthrax, with surprisingly high efficacy towards the bacterium *B. anthracis* [199]. This comparatively simple chemical inhibits growth of *B. anthracis* at several orders of magnitude lower concentrations compared to HU, and whereas mouse RNR was not inhibited at all by the drug, the purified *B. anthracis* RNR was inhibited with several hundred-fold higher potency. Importantly, N-methylhydroxylamine has been shown to be non-toxic to man [235]. With the appearance of multidrug resistant *M. tuberculosis, P. aeruginosa, S. aureus* and several other enterococci, gonococci, streptococci and salmonella, we hypothesize that biochemical characterizations of their RNR enzymes and genetic studies of the strain-specific expression of RNR genes will become increasingly important. Likewise, the unanticipated finding that the severe crop and plant pests *G. zeae (Fusarium graminearum)* and several *Phytophthoras* encode more than one class of RNR genes opens for new promising antiproliferative regimens based on RNR inhibition.

Acknowledgments

We thank Chris Dealwis for sending us high-resolution files for figure 12. Work on RNR in the authors' laboratories was supported by grants from the Ramón y Cajal program and Jeansson Foundations (to ET), and by grants from the Swedish Cancer Foundation and the Swedish Research Council (to BMS).

References

[1] Holmgren A, Reichard P, Thelander L. 1965. Enzymatic synthesis of deoxyribonucleotides, 8. The effects of ATP and dATP in the CDP reductase system from E. coli. *Proc. Natl. Acad. Sci. USA*, 54, 830-836.

[2] Brown NC, Canellakis ZN, Lundin B, Reichard P, Thelander L. 1969. Ribonucleoside diphosphate reductase. Purification of the two subunits, proteins B1 and B2. *Eur. J. Biochem.*, 9, 561-573.

[3] Moore EC, Reichard P. 1964. Enzymatic Synthesis of Deoxyribonucleotides. VI. the Cytidine Diphosphate Reductase System from Novikoff Hepatoma. *J. Biol. Chem.*, 239, 3453-3456.

[4] Blakley RL, Ghambeer RK, Nixon PF, Vitols E. 1965. The cobamide-dependent ribonucleoside triphosphate reductase of lactobacilli. *Biochem. Biophys. Res. Commun.*, 20, 439-445.

[5] Berglund O, Karlström O, Reichard P. 1969. A new ribonucleotide reductase system after infection with phage T4. *Proc. Natl. Acad. Sci. USA*, 62, 829-835.

[6] Carlson J, Fuchs JA, Messing J. 1984. Primary structure of the Escherichia coli ribonucleoside diphosphate reductase operon. *Proc. Natl. Acad. Sci. USA*, 81, 4294-4297.

[7] Caras IW, Levinson BB, Fabry M, Williams SR, Martin DJ. 1985. Cloned mouse ribonucleotide reductase subunit M1 cDNA reveals amino acid sequence homology with Escherichia coli and herpesvirus ribonucleotide reductases. *J. Biol. Chem.*, 260, 7015-7022.

[8] Thelander L, Berg P. 1986. Isolation and characterization of expressible cDNA clones encoding the M1 and M2 subunits of mouse ribonucleotide reductase. *Mol. Cell Biol.*, 6, 3433-3442.

[9] Elledge SJ, Davis RW. 1987. Identification and isolation of the gene encoding the small subunit of ribonucleotide reductase from Saccharomyces cerevisiae: DNA damage-inducible gene required for mitotic viability. *Mol. Cell Biol.*, 7, 2783-2793.
[10] Hurd HK, Roberts CW, Roberts JW. 1987. Identification of the gene for the yeast ribonucleotide reductase small subunit and its inducibility by methyl methanesulfonate. *Mol. Cell Biol.*, 7, 3673-7.
[11] Sartorelli AC, Creasey WA. 1969. Cancer chemotherapy. *Annu. Rev. Pharmacol.*, 9, 51-72.
[12] Krakoff IH, Brown NC, Reichard P. 1968. Inhibition of ribonucleoside diphosphate reductase by hydroxyurea. *Cancer Res.*, 28, 1559-1565.
[13] Atkin CL, Thelander L, Reichard P, Lang G. 1973. Iron and free radical in ribonucleotide reductase. Exchange of iron and Mössbauer spectroscopy of the protein B2 subunit of the Escherichia coli enzyme. *J. Biol. Chem.*, 248, 7464-7472.
[14] Nordlund P, Reichard P. 2006. Ribonucleotide reductases. *Annu. Rev. Biochem.*, 75, 681-706.
[15] Eklund H, Uhlin U, Färnegårdh M, Logan DT, Nordlund P. 2001. Structure and function of the radical enzyme ribonucleotide reductase. *Prog. Biophys. Mol. Biol.*, 77, 177-268.
[16] Chabes A, Thelander L. 2003. DNA building blocks at the foundation of better survival. *Cell Cycle*, 2, 171-173.
[17] Kolberg M, Strand KR, Graff P, Andersson KK. 2004. Structure, function, and mechanism of ribonucleotide reductases. *Biochim. Biophys* Acta, 1699, 1-34.
[18] Stubbe J. 2003. Di-iron-tyrosyl radical ribonucleotide reductases. *Curr. Opin. Chem. Biol.,* 7, 183-188.
[19] Bennati M, Lendzian F, Schmittel M, Zipse H. 2005. Spectroscopic and theoretical approaches for studying radical reactions in class I ribonucleotide reductase. *Biol. Chem.*, 386, 1007-1022.
[20] Sjöberg B-M, Sahlin M. 2002. Thiols in redox mechanism of ribonucleotide reductase. *Methods Enzymol.*, 348, 1-21.
[21] Fontecave M, Mulliez E, Logan DT. 2002. Deoxyribonucleotide synthesis in anaerobic microorganisms: the class III ribonucleotide reductase. *Prog. Nucleic. Acid. Res. Mol. Biol.*, 72, 95-127.
[22] Shao J, Zhou B, Chu B, Yen Y. 2006. Ribonucleotide reductase inhibitors and future drug design. *Curr. Cancer Drug Targets*, 6, 409-431.
[23] Uhlin U, Eklund H. 1994. Structure of ribonucleotide reductase protein R1. *Nature,* 370, 533-539.

[24] Logan DT, Andersson J, Sjöberg B-M, Nordlund P. 1999. A glycyl radical site in the crystal structure of a class III ribonucleotide reductase. *Science*, 283, 1499-1504.

[25] Sintchak MD, Arjara G, Kellogg BA, Stubbe J, Drennan CL. 2002. The crystal structure of class II ribonucleotide reductase reveals how an allosterically regulated monomer mimics a dimer. *Nat. Struct. Biol.*, 9, 293-300.

[26] Kashlan OB, Cooperman BS. 2003. Comprehensive model for allosteric regulation of mammalian ribonucleotide reductase: refinements and consequences. *Biochemistry*, 42, 1696-1706.

[27] Torrents E, Westman M, Sahlin M, Sjöberg B-M. 2006. Ribonucleotide reductase modularity: Atypical duplication of the ATP-cone domain in Pseudomonas aeruginosa. *J. Biol. Chem.*, 281, 25287-25296.

[28] Rofougaran R, Vodnala M, Hofer A. 2006. Enzymatically active mammalian ribonucleotide reductase exists primarily as an alpha6beta2 octamer. *J. Biol. Chem.*, 281, 27705-27711.

[29] Åberg A, Hahne S, Karlsson M, Larsson A, Ormö M, Åhgren A, Sjöberg B-M. 1989. Evidence for two different classes of redox-active cysteines in ribonucleotide reductase of *Escherichia coli*. *J. Biol. Chem.*, 264, 12249-12252.

[30] Mao SS, Holler TP, Yu GX, Bollinger JM, Booker S, Johnston MI, Stubbe J. 1992. A model for the role of multiple cysteine residues involved in ribonucleotide reduction - Amazing and still confusing. *Biochemistry*, 31, 9733-9743.

[31] Petersson L, Gräslund A, Ehrenberg A, Sjöberg B-M, Reichard P. 1980. The iron center in ribonucleotide reductase from Escherichia coli. *J. Biol. Chem.*, 255, 6706-6712.

[32] Larsson A, Sjöberg B-M. 1986. Identification of the stable free radical tyrosine residue in ribonucleotide reductase. *EMBO J.*, 5, 2037-2040.

[33] Nordlund P, Sjöberg B-M, Eklund H. 1990. Three-dimensional structure of the free radical protein of ribonucleotide reductase. *Nature*, 345, 593-598.

[34] Aravind L, Wolf YI, Koonin EV. 2000. The ATP-cone: an evolutionarily mobile, ATP-binding regulatory domain. *J. Mol. Microbiol. Biotechnol.*, 2, 191-194.

[35] Jordan A, Gibert I, Barbe J. 1995. Two different operons for the same function: comparison of the Salmonella typhimurium nrdAB and nrdEF genes. *Gene*, 167, 75-9.

[36] Jordan A, Åslund F, Pontis E, Reichard P, Holmgren A. 1997. Characterization of Escherichia coli NrdH. A glutaredoxin-like protein with a thioredoxin-like activity profile. *J. Biol. Chem.*, 272, 18044-18050.
[37] Larsson K-M, Jordan A, Eliasson R, Reichard P, Logan DT, Nordlund P. 2004. Structural mechanism of allosteric substrate specificity regulation in a ribonucleotide reductase. *Nat. Struct. Mol. Biol.*, 11, 1142-1149.
[38] Tauer A, Benner SA. 1997. The B-12-dependent ribonucleotide reductase from the archaebacterium Thermoplasma acidophila: An evolutionary solution to the ribonucleotide reductase conundrum. *Proc. Natl. Acad. Sci. USA*, 94, 53-58.
[39] Riera J, Robb FT, Weiss R, Fontecave M. 1997. Ribonucleotide reductase in the archaeon Pyrococcus furiosus: A critical enzyme in the evolution of DNA genomes? *Proc. Natl. Acad. Sci USA*, 94, 475-478.
[40] Gleason FK, Olszewski NE. 2002. Isolation of the gene for the B12-dependent ribonucleotide reductase from *Anabaena* sp. strain PCC 7120 and expression in *Escherichia coli*. *J. Bacteriol.*, 184, 6544-6550.
[41] Torrents E, Trevisiol C, Rotte C, Hellman U, Martin W, Reichard P. 2006. Euglena gracilis ribonucleotide reductase: the eukaryote class II enzyme and the possible antiquity of eukaryote B12 dependence. *J. Biol. Chem.*, 281, 5604-5611.
[42] Booker S, Licht S, Broderick J, Stubbe J. 1994. Coenzyme B-12-dependent ribonucleotide reductase: Evidence for the participation of five cysteine residues in ribonucleotide reduction. *Biochemistry*, 33, 12676-12685.
[43] Gerfen GJ, Licht S, Willems JP, Hoffman BM, Stubbe J. 1996. Electron paramagnetic resonance investigations of a kinetically competent intermediate formed in ribonucleotide reduction: Evidence for a thiyl radical-Cob(II)alamin interaction. *J. Am. Chem. Soc.*, 118, 8192-8197.
[44] Torrents E, Eliasson R, Wolpher H, Gräslund A, Reichard P. 2001. The anaerobic ribonucleotide reductase from Lactococcus lactis. Interactions between the two proteins NrdD and NrdG. *J. Biol. Chem.*, 276, 33488-33494.
[45] Andersson J, Westman M, Sahlin M, Sjöberg B-M. 2000. Cysteines involved in radical generation and catalysis of class III anaerobic ribonucleotide reductase. A protein engineering study of bacteriophage T4 NrdD. *J. Biol. Chem.*, 275, 19449-19455.
[46] Young P, Öhman M, Sjöberg B-M. 1994. Bacteriophage T4 gene 55.9 encodes an activity required for anaerobic ribonucleotide reduction. *J. Biol. Chem.*, 269, 27815-27818.

[47] Sun X, Eliasson R, Pontis E, Andersson J, Buist G, Sjöberg B-M, Reichard P. 1995. Generation of the glycyl radical of the anaerobic Escherichia coli ribonucleotide reductase requires a specific activating enzyme. *J. Biol. Chem.*, 270, 2443-2446.

[48] Sofia HJ, Chen G, Hetzler BG, Reyes-Spindola JF, Miller NE. 2001. Radical SAM, a novel protein superfamily linking unresolved steps in familiar biosynthetic pathways with radical mechanisms: functional characterization using new analysis and information visualization methods. *Nucleic. Acids Res.*, 29, 1097-1106.

[49] Tamarit J, Gerez C, Meier C, Mulliez E, Trautwein A, Fontecave M. 2000. The activating component of the anaerobic ribonucleotide reductase from Escherichia coli. An iron-sulfur center with only three cysteines. *J. Biol. Chem.*, 275, 15669-15675.

[50] Torrents E, Poplawski A, Sjöberg B-M. 2005. Two proteins mediate class II ribonucleotide reductase activity in Pseudomonas aeruginosa: expression and transcriptional analysis of the aerobic enzymes. *J. Biol. Chem.*, 280, 16571-16578.

[51] Elledge SJ, Davis RW. 1990. Two genes differentially regulated in the cell cycle and by DNA-damaging agents encode alternative regulatory subunits of ribonucleotide reductase. *Genes Dev.*, 4, 740-51.

[52] Chabes A, Domkin V, Larsson G, Liu A, Gräslund A, Wijmenga S, Thelander L. 2000. Yeast ribonucleotide reductase has a heterodimeric iron-radical-containing subunit. *Proc. Natl. Acad. Sci. USA*, 97, 2474-2479.

[53] Voegtli WC, Ge J, Perlstein DL, Stubbe J, Rosenzweig AC. 2001. Structure of the yeast ribonucleotide reductase Y2Y4 heterodimer. *Proc. Natl. Acad. Sci. USA*, 98, 10073-8.

[54] Håkansson P, Hofer A, Thelander L. 2006. Regulation of mammalian ribonucleotide reduction and dNTP pools after DNA damage and in resting cells. *J. Biol. Chem.*, 281, 7834-7841.

[55] Chabes AL, Björklund S, Thelander L. 2004. S Phase-specific transcription of the mouse ribonucleotide reductase R2 gene requires both a proximal repressive E2F-binding site and an upstream promoter activating region. *J. Biol. Chem.*, 279, 10796-10807.

[56] Tanaka H, Arakawa H, Yamaguchi T, Shiraishi K, Fukuda S, Matsui K, Takei Y, Nakamura Y. 2000. A ribonucleotide reductase gene involved in a p53-dependent cell-cycle checkpoint for DNA damage. *Nature*, 404, 42-49.

[57] Guittet O, Håkansson P, Voevodskaya N, Fridd S, Gräslund A, Arakawa H, Nakamura Y, Thelander L. 2001. Mammalian p53R2 protein forms an active ribonucleotide reductase in vitro with the R1 protein, which is

expressed both in resting cells in response to DNA damage and in proliferating cells. *J. Biol. Chem.*, 276, 40647-40651.

[58] Frame MC, Marsden HS, Dutia BM. 1985. The ribonucleotide reductase induced by herpes simplex virus type 1 involves minimally a complex of two polypeptides (136K and 38K). *J. Gen. Virol.*, 66, 1581-1587.

[59] Dolan A, Jamieson FE, Cunningham C, Barnett BC, McGeoch DJ. 1998. The genome sequence of herpes simplex virus type 2. *J. Virol.*, 72, 2010-2021.

[60] Andersson J, Westman M, Hofer A, Sjöberg B-M. 2000. Allosteric regulation of the class III anaerobic ribonucleotide reductase from bacteriophage T4. *J. Biol. Chem.*, 275, 19443-19448.

[61] Berglund O. 1975. Ribonucleoside diphosphate reductase induced by bacteriophage T4. III. Isolation and characterization of proteins B1 and B2. *J. Biol. Chem.*, 250, 7450-7455.

[62] Roshick C, Iliffe-Lee ER, McClarty G. 2000. Cloning and characterization of ribonucleotide reductase from *Chlamydia trachomatis*. *J. Biol. Chem.*, 275, 38111-38119.

[63] Bateman A, Coin L, Durbin R, Finn RD, Hollich V, Griffiths-Jones S, Khanna A, Marshall M, Moxon S, Sonnhammer EL, Studholme DJ, Yeats C, Eddy SR. 2004. The Pfam protein families database. *Nucleic. Acids Res.*, 32, D138-D141.

[64] Friedrich NC, Torrents E, Gibb EA, Sahlin M, Sjöberg B-M, Edgell DR. 2007. Insertion of a homing endonuclease creates a genes-in-pieces ribonucleotide reductase that retains function. *Proc. Natl. Acad. Sci. USA*, in press.

[65] Logan DT, Mulliez E, Larsson KM, Bodevin S, Atta M, Garnaud PE, Sjöberg B-M, Fontecave M. 2003. A metal-binding site in the catalytic subunit of anaerobic ribonucleotide reductase. *Proc. Natl. Acad. Sci. USA*, 100, 3826-3831.

[66] Siedow A, Cramm R, Siddiqui RA, Friedrich B. 1999. A megaplasmid-borne anaerobic ribonucleotide reductase in Alcaligenes eutrophus H16. *J. Bacteriol.*, 181, 4919-4928.

[67] Schwartz E, Henne A, Cramm R, Eitinger T, Friedrich B, Gottschalk G. 2003. Complete nucleotide sequence of pHG1: a Ralstonia eutropha H16 megaplasmid encoding key enzymes of H(2)-based ithoautotrophy and anaerobiosis. *J. Mol. Biol.*, 332, 369-383.

[68] Pohlmann A, Fricke WF, Reinecke F, Kusian B, Liesegang H, Cramm R, Eitinger T, Ewering C, Potter M, Schwartz E, Strittmatter A, Voss I, Gottschalk G, Steinbuchel A, Friedrich B, Bowien B. 2006. Genome

sequence of the bioplastic-producing "Knallgas" bacterium Ralstonia eutropha H16. *Nat. Biotechnol.*, 24, 1257-1262.

[69] Högbom M, Stenmark P, Voevodskaya N, McClarty G, Gräslund A, Nordlund P. 2004. The radical site in chlamydial ribonucleotide reductase defines a new R2 subclass. *Science*, 305, 245-248.

[70] Voevodskaya N, Narvaez AJ, Domkin V, Torrents E, Thelander L, Gräslund A. 2006. Chlamydial ribonucleotide reductase: tyrosyl radical function in catalysis replaced by the FeIII-FeIV cluster. *Proc. Natl. Acad. Sci. USA*, 103, 9850-9854.

[71] Strand KR, Karlsen S, Kolberg M, Rohr AK, Gorbitz CH, Andersson KK. 2004. Crystal structural studies of changes in the native dinuclear iron center of ribonucleotide reductase protein R2 from mouse. *J. Biol. Chem.*, 279, 46794-46801.

[72] Eriksson M, Jordan A, Eklund H. 1998. Structure of Salmonella typhimurium nrdF ribonucleotide reductase in its oxidized and reduced forms. *Biochemistry*, 37, 13359-13369.

[73] Högbom M, Huque Y, Sjöberg B-M, Nordlund P. 2002. Crystal structure of the di-iron/radical protein of ribonucleotide reductase from *Corynebacterium ammoniagenes*. *Biochemistry*, 41, 1381-1389.

[74] Hutchison CA, Peterson SN, Gill SR, Cline RT, White O, Fraser CM, Smith HO, Venter JC. 1999. Global transposon mutagenesis and a minimal Mycoplasma genome. *Science*, 286, 2165-2169.

[75] Poole AM, Logan DT, Sjöberg B-M. 2002. The evolution of the ribonucleotide reductases: much ado about oxygen. *J. Mol. Evol.*, 55, 180-196.

[76] Torrents E, Aloy P, Gibert I, Rodriguez-Trelles F. 2002. Ribonucleotide reductases: divergent evolution of an ancient enzyme. *J. Mol. Evol.*, 55, 138-152.

[77] Stubbe J. 2000. Ribonucleotide reductases: the link between an RNA and a DNA world? *Curr. Opin. Struct. Biol.*, 10, 731-736.

[78] Reichard P. 1993. From RNA to DNA, why so many ribonucleotide reductases? *Science*, 260, 1773-1777.

[79] Reichard P. 1997. The evolution of ribonucleotide reduction. *Trends Biochem. Sci.*, 22, 81-85.

[80] Reichard P. 2002. Ribonucleotide reductases: the evolution of allosteric regulation. *Arch. Biochem. Biophys.*, 397, 149-55.

[81] Fraser HB. 2005. Modularity and evolutionary constraint on proteins. *Nat. Genet.*, 37, 351-352.

[82] Nikitin F, Lisacek F. 2003. Investigating protein domain combinations in complete proteomes. *Comput. Biol. Chem.*, 27, 481-495.
[83] Uhlin U, Eklund H. 1996. The ten-stranded β/α barrel in ribonucleotide reductase protein R1. *J. Mol. Biol.*, 262, 358-369.
[84] Selmer T, Pierik AJ, Heider J. 2005. New glycyl radical enzymes catalysing key metabolic steps in anaerobic bacteria. *Biol. Chem.*, 386, 981-988.
[85] Uppsten M, Färnegårdh M, Jordan A, Eliasson R, Eklund H, Uhlin U. 2003. Structure of the large subunit of class Ib ribonucleotide reductase from Salmonella typhimurium and its complexes with allosteric effectors. *J. Mol. Biol.*, 330, 87-97.
[86] Holmgren A. 1989. Thioredoxin and glutaredoxin systems. *J. Biol. Chem.*, 264, 13963-13966.
[87] McCready S, Muller JA, Boubriak I, Berquist BR, Ng WL, Dassarma S. 2005. UV irradiation induces homologous recombination genes in the model archaeon, Halobacterium sp. NRC-1. Saline Systems, 1, 3.
[88] Torrents E, Roca I, Gibert I. 2004. The open reading frame present in the nrdEF cluster of Corynebacterium ammoniagenes is a transcriptional regulator belonging to the GntR family. *Curr. Microbiol.*, 49, 152-157.
[89] Torrents E, Roca I, Gibert I. 2003. Corynebacterium ammoniagenes class Ib ribonucleotide reductase: transcriptional regulation of an atypical genomic organization in the nrd cluster. *Microbiology*, 149, 1011-1020.
[90] Torrents E, Jordan A, Karlsson M, Gibert I. 2000. Occurrence of multiple ribonucleotide reductase classes in gamma-proteobacteria species. *Curr. Microbiol.*, 41, 346-351.
[91] Sun X, Harder J, Krook M, Jörnvall H, Sjöberg B-M, Reichard P. 1993. A possible glycine radical in anaerobic ribonucleotide reductase from Escherichia coli: nucleotide sequence of the cloned nrdD gene. *Proc. Natl. Acad. Sci. USA*, 90, 577-581.
[92] Augustin LB, Jacobson BA, Fuchs JA. 1994. Escherichia coli Fis and DnaA proteins bind specifically to the nrd promoter region and affect expression of an nrd-lac fusion. *J. Bacteriol.*, 176, 378-387.
[93] Jordan A, Gibert I, Barbé J. 1994. Cloning and sequencing of the genes from Salmonella typhimurium encoding a new bacterial ribonucleotide reductase. *J. Bacteriol.*, 176, 3420-3427.
[94] Blattner FR, Plunkett G, 3rd, Bloch CA, Perna NT, Burland V, Riley M, Collado-Vides J, Glasner JD, Rode CK, Mayhew GF, Gregor J, Davis NW, Kirkpatrick HA, Goeden MA, Rose DJ, Mau B, Shao Y. 1997. The complete genome sequence of Escherichia coli K-12. *Science*, 277, 1453-1474.

[95] Jacobson BA, Fuchs JA. 1998. A 45 bp inverted repeat is required for cell cycle regulation of the Escherichia coli nrd operon. *Mol. Microbiol.*, 28, 1307-1314.

[96] Filpula D, Fuchs JA. 1977. Regulation of ribonucleoside diphosphate reductase synthesis in Escherichia coli: increased enzyme synthesis as a result of inhibition of deoxyribonucleic acid synthesis. *J. Bacteriol.*, 130, 107-113.

[97] Peter BJ, Arsuaga J, Breier AM, Khodursky AB, Brown PO, Cozzarelli NR. 2004. Genomic transcriptional response to loss of chromosomal supercoiling in Escherichia coli. *Genome Biol.*, 5, R87.

[98] Dwyer DJ, Kohanski MA, Hayete B, Collins JJ. 2007. Gyrase inhibitors induce an oxidative damage cellular death pathway in Escherichia coli. *Mol. Syst. Biol.*, 3, 91.

[99] Han JS, Kwon HS, Yim JB, Hwang DS. 1998. Effect of IciA protein on the expression of the nrd gene encoding ribonucleoside diphosphate reductase in E. coli. *Mol. Gen. Genet*, 259, 610-614.

[100] Gon S, Camara JE, Klungsoyr HK, Crooke E, Skarstad K, Beckwith J. 2006. A novel regulatory mechanism couples deoxyribonucleotide synthesis and DNA replication in Escherichia coli. *Embo J.*, 25, 1137-1147.

[101] Gallardo-Madueno R, Leal JF, Dorado G, Holmgren A, Lopez-Barea J, Pueyo C. 1998. In vivo transcription of nrdAB operon and of grxA and fpg genes is triggered in Escherichia coli lacking both thioredoxin and glutaredoxin 1 or thioredoxin and glutathione, respectively. *J. Biol. Chem.*, 273, 18382-18388.

[102] Ortenberg R, Gon S, Porat A, Beckwith J. 2004. Interactions of glutaredoxins, ribonucleotide reductase, and components of the DNA replication system of Escherichia coli. *Proc. Natl. Acad. Sci. USA*, 101, 7439-7444.

[103] Gibert I, Calero S, Barbé J. 1990. Measurement of in vivo expression of nrdA and nrdB genes of Escherichia coli by using lacZ gene fusions. *Mol. Gen. Genet.*, 220, 400-408.

[104] Courcelle J, Khodursky A, Peter B, Brown PO, Hanawalt PC. 2001. Comparative gene expression profiles following UV exposure in wild-type and SOS-deficient Escherichia coli. *Genetics*, 158, 41-64.

[105] Jordan A, Aragall E, Gibert I, Barbé J. 1996. Promoter identification and expression analysis of Salmonella typhimurium and Escherichia coli nrdEF operons encoding one of two class I ribonucleotide reductases present in both bacteria. *Mol. Microbiol.*, 19, 777-790.

[106] Monje-Casas F, Jurado J, Prieto-Alamo MJ, Holmgren A, Pueyo C. 2001. Expression analysis of the nrdHIEF operon from Escherichia coli. Conditions that trigger the transcript level in vivo. *J. Biol. Chem.*, 276, 18031-18037.
[107] Zheng M, Wang X, Templeton LJ, Smulski DR, LaRossa RA, Storz G. 2001. DNA microarray-mediated transcriptional profiling of the Escherichia coli response to hydrogen peroxide. *J. Bacteriol.*, 183, 4562-4570.
[108] Mukhopadhyay P, Zheng M, Bedzyk LA, LaRossa RA, Storz G. 2004. Prominent roles of the NorR and Fur regulators in the Escherichia coli transcriptional response to reactive nitrogen species. *Proc. Natl. Acad. Sci. USA*, 101, 745-750.
[109] Justino MC, Vicente JB, Teixeira M, Saraiva LM. 2005. New genes implicated in the protection of anaerobically grown Escherichia coli against nitric oxide. *J. Biol. Chem.*, 280, 2636-2643.
[110] Flatley J, Barrett J, Pullan ST, Hughes MN, Green J, Poole RK. 2005. Transcriptional responses of Escherichia coli to S-nitrosoglutathione under defined chemostat conditions reveal major changes in methionine biosynthesis. *J. Biol. Chem.*, 280, 10065-10072.
[111] Vassinova N, Kozyrev D. 2000. A method for direct cloning of fur-regulated genes: identification of seven new fur-regulated loci in Escherichia coli. *Microbiology,* 146 Pt 12, 3171-3182.
[112] McHugh JP, Rodriguez-Quinones F, Abdul-Tehrani H, Svistunenko DA, Poole RK, Cooper CE, Andrews SC. 2003. Global iron-dependent gene regulation in Escherichia coli. A new mechanism for iron homeostasis. *J. Biol. Chem.*, 278, 29478-29486.
[113] Zhou D, Qin L, Han Y, Qiu J, Chen Z, Li B, Song Y, Wang J, Guo Z, Zhai J, Du Z, Wang X, Yang R. 2006. Global analysis of iron assimilation and fur regulation in Yersinia pestis. *FEMS Microbiol. Lett.*, 258, 9-17.
[114] Boston T, Atlung T. 2003. FNR-mediated oxygen-responsive regulation of the nrdDG operon of Escherichia coli. *J. Bacteriol.*, 185, 5310-5313.
[115] Garriga X, Eliasson R, Torrents E, Jordan A, Barbé J, Gibert I, Reichard P. 1996. nrdD and nrdG genes are essential for strict anaerobic growth of Escherichia coli. *Biochem. Biophys. Res. Commun.*, 229, 189-192.
[116] Casado C, Llagostera M, Barbé J. 1991. Expression of nrdA and nrdB genes of Escherichia coli is decreased under anaerobiosis. *FEMS Microbiol. Lett.*, 67, 153-157.
[117] Kang Y, Weber KD, Qiu Y, Kiley PJ, Blattner FR. 2005. Genome-wide expression analysis indicates that FNR of Escherichia coli K-12 regulates a large number of genes of unknown function. *J. Bacteriol.*, 187, 1135-1160.

[118] Faucher SP, Porwollik S, Dozois CM, McClelland M, Daigle F. 2006. Transcriptome of Salmonella enterica serovar Typhi within macrophages revealed through the selective capture of transcribed sequences. *Proc. Natl. Acad. Sci. USA*, 103, 1906-1911.

[119] Jordan A, Torrents E, Sala I, Hellman U, Gibert I, Reichard P. 1999. Ribonucleotide reduction in Pseudomonas species: simultaneous presence of active enzymes from different classes. *J. Bacteriol.*, 181, 3974-3980.

[120] Vollack KU, Härtig E, Korner H, Zumft WG. 1999. Multiple transcription factors of the FNR family in denitrifying Pseudomonas stutzeri: characterization of four fnr-like genes, regulatory responses and cognate metabolic processes. *Mol. Microbiol.*, 31, 1681-1694.

[121] Lazarevic V, Soldo B, Dusterhoft A, Hilbert H, Mauel C, Karamata D. 1998. Introns and intein coding sequence in the ribonucleotide reductase genes of Bacillus subtilis temperate bacteriophage SPbeta. *Proc. Natl. Acad. Sci. USA*, 95, 1692-1697.

[122] Lazarevic V, Dusterhoft A, Soldo B, Hilbert H, Mauel C, Karamata D. 1999. Nucleotide sequence of the Bacillus subtilis temperate bacteriophage SPbetac2. *Microbiology*, 145 (Pt 5), 1055-1067.

[123] Härtig E, Hartmann A, Schätzle M, Albertini AM, Jahn D. 2006. The Bacillus subtilis nrdEF genes, encoding a class Ib ribonucleotide reductase, are essential for aerobic and anaerobic growth. *Appl. Environ. Microbiol.*, 72, 5260-5265.

[124] Yellaboina S, Ranjan S, Chakhaiyar P, Hasnain SE, Ranjan A. 2004. Prediction of DtxR regulon: identification of binding sites and operons controlled by Diphtheria toxin repressor in Corynebacterium diphtheriae. *BMC Microbiol.*, 4, 38.

[125] Yang F, Curran SC, Li LS, Avarbock D, Graf JD, Chua MM, Lu G, Salem J, Rubin H. 1997. Characterization of two genes encoding the Mycobacterium tuberculosis ribonucleotide reductase small subunit. *J. Bacteriol.*, 179, 6408-6415.

[126] Dawes SS, Warner DF, Tsenova L, Timm J, McKinney JD, Kaplan G, Rubin H, Mizrahi V. 2003. Ribonucleotide reduction in Mycobacterium tuberculosis: function and expression of genes encoding class Ib and class II ribonucleotide reductases. *Infect. Immun.*, 71, 6124-6131.

[127] Masalha M, Borovok I, Schreiber R, Aharonowitz Y, Cohen G. 2001. Analysis of transcription of the Staphylococcus aureus aerobic class Ib and anaerobic class III ribonucleotide reductase genes in response to oxygen. *J. Bacteriol.*, 183, 7260-7272.

[128] Borovok I, Gorovitz B, Yanku M, Schreiber R, Gust B, Chater K, Aharonowitz Y, Cohen G. 2004. Alternative oxygen-dependent and oxygen-independent ribonucleotide reductases in Streptomyces: cross-regulation and physiological role in response to oxygen limitation. *Mol. Microbiol.*, 54, 1022-1035.

[129] Borovok I, Kreisberg-Zakarin R, Yanko M, Schreiber R, Myslovati M, Åslund F, Holmgren A, Cohen G, Aharonowitz Y. 2002. Streptomyces spp. contain class Ia and class II ribonucleotide reductases: expression analysis of the genes in vegetative growth. *Microbiology*, 148, 391-404.

[130] Vitreschak AG, Rodionov DA, Mironov AA, Gelfand MS. 2003. Regulation of the vitamin B12 metabolism and transport in bacteria by a conserved RNA structural element. *RNA*, 9, 1084-1097.

[131] Vitreschak AG, Rodionov DA, Mironov AA, Gelfand MS. 2004. Riboswitches: the oldest mechanism for the regulation of gene expression? *Trends Genet*, 20, 44-50.

[132] Winkler WC. 2005. Riboswitches and the role of noncoding RNAs in bacterial metabolic control. *Curr. Opin. Chem. Biol.*, 9, 594-602.

[133] Gilbert SD, Batey RT. 2006. Riboswitches: fold and function. *Chem. Biol.*, 13, 805-807.

[134] Borovok I, Gorovitz B, Schreiber R, Aharonowitz Y, Cohen G. 2006. Coenzyme B12 controls transcription of the Streptomyces class Ia ribonucleotide reductase nrdABS operon via a riboswitch mechanism. *J. Bacteriol.*, 188, 2512-2520.

[135] Rodionov DA, Gelfand MS. 2005. Identification of a bacterial regulatory system for ribonucleotide reductases by phylogenetic profiling. *Trends Genet*, 21, 385-389.

[136] Grinberg I, Shteinberg T, Gorovitz B, Aharonowitz Y, Cohen G, Borovok I. 2006. The Streptomyces NrdR transcriptional regulator is a Zn ribbon/ATP cone protein that binds to the promoter regions of class Ia and class II ribonucleotide reductase operons. *J. Bacteriol.*, 188, 7635-7644.

[137] Torrents E, Grinberg I, Gorovitz B, Lundström H, Borovok I, Aharonowitz Y, Sjöberg B-M, Cohen G. 2007. NrdR controls differential expression of the Escherichia coli ribonucleotide reductase genes. *Submitted to J. Bacteriol.*

[138] Wang PJ, Chabes A, Casagrande R, Tian XC, Thelander L, Huffaker TC. 1997. Rnr4p, a novel ribonucleotide reductase small-subunit protein. *Mol. Cell Biol.*, 17, 6114-21.

[139] Huang M, Elledge SJ. 1997. Identification of RNR4, encoding a second essential small subunit of ribonucleotide reductase in Saccharomyces cerevisiae. *Mol. Cell Biol.*, 17, 6105-13.
[140] Chabes A, Domkin V, Thelander L. 1999. Yeast Sml1, a protein inhibitor of ribonucleotide reductase. *J. Biol. Chem.*, 274, 36679-83.
[141] Zhao X, Chabes A, Domkin V, Thelander L, Rothstein R. 2001. The ribonucleotide reductase inhibitor Sml1 is a new target of the Mec1/Rad53 kinase cascade during growth and in response to DNA damage. *Embo J.*, 20, 3544-53.
[142] Lee YD, Elledge SJ. 2006. Control of ribonucleotide reductase localization through an anchoring mechanism involving Wtm1. *Genes Dev.*, 20, 334-44.
[143] Wechsler JA, Gross JD. 1971. Escherichia coli mutants temperature-sensitive for DNA synthesis. *Mol. Gen. Genet.*, 113, 273-284.
[144] Fuchs JA, Karlström HO, Warner HR, Reichard P. 1972. Defective gene product in *dnaF* mutant of *Escherichia coli*. Nature New Biol, 238, 69-71.
[145] Fuchs JA, Karlström HO. 1973. A mutant of Escherichia coli defective in ribonucleosidediphosphate reductase. 2. Characterization of the enzymatic defect. *Eur. J. Biochem.*, 32, 457-462.
[146] Fuchs JA, Neuhard J. 1973. A mutant of Escherichia coli defective in ribonucleosidediphosphate reductase. 1. Isolation of the mutant as a deoxyuridine auxotroph. *Eur. J. Biochem.*, 32, 451-456.
[147] Kren B, Fuchs JA. 1987. Characterization of the *ftsB* gene as an allele of the *nrdB* gene in *Escherichia coli*. *J. Bacteriol.*, 169, 14-8.
[148] Taschner PE, Verest JG, Woldringh CL. 1987. Genetic and morphological characterization of ftsB and nrdB mutants of Escherichia coli. *J. Bacteriol.*, 169, 19-25.
[149] Larsson A, Climent I, Nordlund P, Sahlin M, Sjöberg B-M. 1996. Structural and functional characterization of two mutated R2 proteins of Escherichia coli ribonucleotide reductase. *Eur. J. Biochem.*, 237, 58-63.
[150] Caras IW, Martin DWJ. 1988. Molecular cloning of the cDNA for a mutant mouse ribonucleotide reductase M1 that produces a dominant mutator phenotype in mammalian cells. *Mol. Cell Biol.*, 8, 2698-2704.
[151] Smeds J, Kumar R, Hemminki K. 2001. Polymorphic insertion of additional repeat within an area of direct 8 bp tandem repeats in the 5'-untranslated region of the p53R2 gene and cancer risk. *Mutagenesis*, 16, 547-550.
[152] Byun DS, Chae KS, Ryu BK, Lee MG, Chi SG. 2002. Expression and mutation analyses of P53R2, a newly identified p53 target for DNA repair in human gastric carcinoma. *Int. J. Cancer*, 98, 718-723.

[153] Ye Z, Parry JM. 2002. The discovery and confirmation of single nucleotide polymorphisms in the human p53R2 gene by EST database analysis. *Mutagenesis*, 17, 361-364.
[154] Hayashi H, Furihata M, Kuwahara M, Kagawa S, Shuin T, Ohtsuki Y. 2004. Infrequent alteration in the p53R2 gene in human transitional cell carcinoma of the urinary tract. *Pathobiology*, 71, 103-106.
[155] Deng ZL, Xie DW, Bostick RM, Miao XJ, Gong YL, Zhang JH, Wargovich MJ. 2005. Novel genetic variations of the p53R2 gene in patients with colorectal adenoma and controls. *World J. Gastroenterol.*, 11, 5169-5173.
[156] Kwon WS, Rha SY, Choi YH, Lee JO, Park KH, Jung JJ, Kim TS, Jeung HC, Chung HC. 2006. Ribonucleotide reductase M1 (RRM1) 2464G>A polymorphism shows an association with gemcitabine chemosensitivity in cancer cell lines. *Pharmacogenet Genomics*, 16, 429-438.
[157] Sjöberg B-M, Eklund H, Fuchs JA, Carlson J, Standart NM, Ruderman JV, Bray SJ, Hunt T. 1985. Identification of the stable free radical tyrosine residue in ribonucleotide reductase. A sequence comparison. *Febs Lett.*, 183, 99-102.
[158] Sun XY, Ollagnier S, Schmidt PP, Atta M, Mulliez E, Lepape L, Eliasson R, Gräslund A, Fontecave M, Reichard P, Sjöberg B-M. 1996. The free radical of the anaerobic ribonucleotide reductase from *Escherichia coli* is at glycine 681. *J. Biol. Chem.*, 271, 6827-6831.
[159] Young P, Andersson J, Sahlin M, Sjöberg B-M. 1996. Bacteriophage T4 anaerobic ribonucleotide reductase contains a stable glycyl radical at position 580. *J. Biol. Chem.*, 271, 20770-20775.
[160] Kolberg M, Logan DT, Bleifuss G, Pötsch S, Sjöberg B-M, Gräslund A, Lubitz W, Lassmann G, Lendzian F. 2005. A new tyrosyl radical on Phe208 as ligand to the diiron center in Escherichia coli ribonucleotide reductase, mutant R2-Y122H. Combined x-ray diffraction and EPR/ENDOR studies. *J. Biol. Chem.*, 280, 11233-11246.
[161] Pötsch S, Lendzian F, Ingemarson R, Hornberg A, Thelander L, Lubitz W, Lassmann G, Gräslund A. 1999. The iron-oxygen reconstitution reaction in protein R2-tyr-177 mutants of mouse ribonucleotide reductase - EPR and electron nuclear double resonance studies on a new transient tryptophan radical. *J. Biol. Chem.*, 274, 17696-17704.
[162] Adrait A, Öhrström M, Barra AL, Thelander L, Gräslund A. 2002. EPR studies on a stable sulfinyl radical observed in the iron-oxygen-reconstituted Y177F/I263C protein R2 double mutant of ribonucleotide reductase from mouse. *Biochemistry*, 41, 6510-6516.

[163] Persson BO, Karlsson M, Climent I, Ling JS, Sanders Loehr J, Sahlin M, Sjöberg B-M. 1996. Iron ligand mutants in protein R2 of *Escherichia coli* ribonucleotide reductase - Retention of diiron site, tyrosyl radical and enzymatic activity in mutant proteins lacking an iron-binding side chain. *J. Biol. Inorg. Chem.*, 1, 247-256.

[164] Voegtli WC, Sommerhalter M, Saleh L, Baldwin J, Bollinger JM, Jr., Rosenzweig AC. 2003. Variable coordination geometries at the diiron(II) active site of ribonucleotide reductase R2. *J. Am. Chem. Soc.*, 125, 15822-15830.

[165] Ormö M, Regnström K, Wang ZG, Que L, Sahlin M, Sjöberg B-M. 1995. Residues important for radical stability in ribonucleotide reductase from *Escherichia coli*. *J. Biol. Chem.*, 270, 6570-6576.

[166] Climent I, Sjöberg B-M, Huang CY. 1992. Site-directed mutagenesis and deletion of the carboxyl terminus of *Escherichia coli* ribonucleotide reductase protein R2. Effects on catalytic activity and subunit interaction. *Biochemistry*, 31, 4801-4807.

[167] Ekberg M, Pötsch S, Sandin E, Thunnissen M, Nordlund P, Sahlin M, Sjöberg B-M. 1998. Preserved catalytic activity in an engineered ribonucleotide reductase R2 protein with a nonphysiological radical transfer pathway - The importance of hydrogen bond connections between the participating residues. *J. Biol. Chem.*, 273, 21003-21008.

[168] Tong W, Burdi D, RiggsGelasco P, Chen S, Edmondson D, Huynh BH, Stubbe J, Han S, Arvai A, Tainer J. 1998. Characterization of Y122F R2 of Escherichia coli ribonucleotide reductase by time-resolved physical biochemical methods and X-ray crystallography. *Biochemistry*, 37, 5840-5848.

[169] Parkin SE, Chen SX, Ley BA, Mangravite L, Edmondson DE, Huynh BH, Bollinger JM. 1998. Electron injection through a specific pathway determines the outcome of oxygen activation at the diiron cluster in the F208Y mutant of Escherichia coli ribonucleotide reductase protein R2. *Biochemistry*, 37, 1124-1130.

[170] Sahlin M, Cho KB, Pötsch S, Lytton SD, Huque Y, Gunther MR, Sjöberg B-M, Mason RP, Gräslund A. 2002. Peroxyl adduct radicals formed in the iron/oxygen reconstitution reaction of mutant ribonucleotide reductase R2 proteins from *Escherichia coli*. *J. Biol. Inorg. Chem.*, 7, 74-82.

[171] Baldwin J, Krebs C, Saleh L, Stelling M, Huynh BH, Bollinger JM, Jr., Riggs-Gelasco P. 2003. Structural characterization of the peroxodiiron(III) intermediate generated during oxygen activation by the W48A/D84E

variant of ribonucleotide reductase protein R2 from Escherichia coli. *Biochemistry*, 42, 13269-13279.
[172] Rova U, Adrait A, Pötsch S, Gräslund A, Thelander L. 1999. Evidence by mutagenesis that Tyr(370) of the mouse ribonucleotide reductase R2 protein is the connecting link in the intersubunit radical transfer pathway. *J. Biol. Chem.*, 274, 23746-23751.
[173] Rova U, Goodtzova K, Ingemarson R, Behravan G, Gräslund A, Thelander L. 1995. Evidence by site-directed mutagenesis supports long-range electron transfer in mouse ribonucleotide reductase. *Biochemistry*, 34, 4267-4275.
[174] Schmidt PP, Rova U, Katterle B, Thelander L, Gräslund A. 1998. Kinetic evidence that a radical transfer pathway in protein R2 of mouse ribonucleotide reductase is involved in generation of the tyrosyl free radical. *J. Biol. Chem.*, 273, 21463-21472.
[175] Seyedsayamdost MR, Yee CS, Reece SY, Nocera DG, Stubbe J. 2006. pH Rate profiles of FnY356-R2s (n = 2, 3, 4) in Escherichia coli ribonucleotide reductase: evidence that Y356 is a redox-active amino acid along the radical propagation pathway. *J. Am. Chem. Soc.*, 128, 1562-1568.
[176] Thelander L, Larsson B. 1976. Active site of ribonucleoside diphosphate reductase from *Escherichia coli*. Inactivation of the enzyme by 2'-substituted ribonucleoside diphosphates. *J. Biol. Chem.*, 251, 1398-1405.
[177] Sjöberg B-M, Gräslund A, Eckstein F. 1983. A substrate radical intermediate in the reaction between ribonucleotide reductase from *Escherichia coli* and 2'-azido-2'-deoxynucleoside diphosphates. *J. Biol. Chem.*, 258, 8060-8067.
[178] Ator MA, Stubbe J. 1985. Mechanism of inactivation of Escherichia coli ribonucleotide reductase by 2'-chloro-2'-deoxyuridine 5'-diphosphate: evidence for generation of a 2'-deoxy-3'-ketonucleotide via a net 1,2 hydrogen shift. *Biochemistry*, 24, 7214-7221.
[179] Fritscher J, Artin E, Wnuk S, Bar G, Robblee JH, Kacprzak S, Kaupp M, Griffin RG, Bennati M, Stubbe J. 2005. Structure of the nitrogen-centered radical formed during inactivation of E. coli ribonucleotide reductase by 2'-azido-2'-deoxyuridine-5'-diphosphate: trapping of the 3'-ketonucleotide. *J. Am. Chem. Soc.*, 127, 7729-7738.
[180] van der Donk WA, Yu GX, Perez L, Sanchez RJ, Stubbe J, Samano V, Robins MJ. 1998. Detection of a new substrate-derived radical during inactivation of ribonucleotide reductase from Escherichia coli by gemcitabine 5'-diphosphate. *Biochemistry*, 37, 6419-6426.

[181] Gallicchio VS. 2005. Ribonucleotide reductase: target therapy for human disease. *Expert Opinion on Therapeutic Patents,* 15, 659-673.
[182] Ashley GW, Harris G, Stubbe J. 1988. Inactivation of the Lactobacillus leichmannii ribonucleoside triphosphate reductase by 2'-chloro-2'-deoxyuridine 5'-triphosphate: stoichiometry of inactivation, site of inactivation, and mechanism of the protein chromophore formation. *Biochemistry,* 27, 4305-4310.
[183] Silva DJ, Stubbe J, Samano V, Robins MJ. 1998. Gemcitabine 5'-triphosphate is a stoichiometric mechanism-based inhibitor of Lactobacillus leichmannii ribonucleoside triphosphate reductase: Evidence for thiyl radical-mediated nucleotide radical formation. *Biochemistry,* 37, 5528-5535.
[184] Davidson JD, Ma L, Flagella M, Geeganage S, Gelbert LM, Slapak CA. 2004. An increase in the expression of ribonucleotide reductase large subunit 1 is associated with gemcitabine resistance in non-small cell lung cancer cell lines. *Cancer Res.,* 64, 3761-3766.
[185] Wright JA, Alam TG, McClarty GA, Tagger AY, Thelander L. 1987. Altered expression of ribonucleotide reductase and role of M2 gene amplification in hydroxyurea-resistant hamster, mouse, rat, and human cell lines. *Somat. Cell Mol. Genet.,* 13, 155-165.
[186] Hurta RA, Wright JA. 1990. Amplification of the genes for both components of ribonucleotide reductase in hydroxyurea resistant mammalian cells. *Biochem. Biophys. Res. Commun.,* 167, 258-264.
[187] Goan YG, Zhou BS, Hu E, Mi S, Yen Y. 1999. Overexpression of ribonucleotide reductase as a mechanism of resistance to 2,2-difluorodeoxycytidine in the human KB cancer cell line. Cancer Res, 59, 4204-4207.
[188] Xu H, Faber C, Uchiki T, Racca J, Dealwis C. 2006. Structures of eukaryotic ribonucleotide reductase I define gemcitabine diphosphate binding and subunit assembly. *Proc. Natl. Acad. Sci. USA,* 103, 4028-4033.
[189] Harris G, Ator M, Stubbe J. 1984. Mechanism of inactivation of Escherichia coli and Lactobacillus leichmannii ribonucleotide reductases by 2'-chloro-2'-deoxynucleotides: evidence for generation of 2-methylene-3(2H)-furanone. *Biochemistry,* 23, 5214-5225.
[190] Pereira S, Fernandes PA, Ramos MJ. 2004. Mechanism for ribonucleotide reductase inactivation by the anticancer drug gemcitabine. *J. Comput. Chem.,* 25, 1286-1294.

[191] Heinemann V, Xu YZ, Chubb S, Sen A, Hertel LW, Grindey GB, Plunkett W. 1990. Inhibition of ribonucleotide reduction in CCRF-CEM cells by 2',2'-difluorodeoxycytidine. *Mol. Pharmacol.*, 38, 567-572.
[192] Jordheim LP, Guittet O, Lepoivre M, Galmarini CM, Dumontet C. 2005. Increased expression of the large subunit of ribonucleotide reductase is involved in resistance to gemcitabine in human mammary adenocarcinoma cells. *Mol. Cancer Ther.*, 4, 1268-1276.
[193] Jordheim LP, Galmarini CM, Dumontet C. 2006. Gemcitabine resistance due to deoxycytidine kinase deficiency can be reverted by fruitfly deoxynucleoside kinase, DmdNK, in human uterine sarcoma cells. *Cancer Chemother. Pharmacol.*, 58, 547-554.
[194] Kanazawa J, Takahashi T, Akinaga S, Tamaoki T, Okabe M. 1998. The relationship between the antitumor activity and the ribonucleotide reductase inhibitory activity of (E)-2'-deoxy-2'-(fluoromethylene) cytidine, MDL 101,731. *Anticancer Drugs*, 9, 653-657.
[195] van der Donk WA, Yu GX, Silva DJ, Stubbe J. 1996. Inactivation of ribonucleotide reductase by (E)2'-fluoromethylene-2'-deoxycytidine 5'-diphosphate: A paradigm for nucleotide mechanism-based. *Biochemistry*, 35, 8381-8391.
[196] Masuda N, Matsui K, Yamamoto N, Nogami T, Nakagawa K, Negoro S, Takeda K, Takifuji N, Yamada M, Kudoh S, Okuda T, Nemoto S, Ogawa K, Myobudani H, Nihira S, Fukuoka M. 2000. Phase I trial of oral 2'-deoxy-2'-methylidenecytidine: on a daily x 14-day schedule. *Clin. Cancer Res.*, 6, 2288-2294.
[197] Baker CH, Banzon J, Bollinger JM, Stubbe J, Samano V, Robins MJ, Lippert B, Jarvi E, Resvick R. 1991. 2'-Deoxy-2'-methylenecytidine and 2'-deoxy-2',2'-difluorocytidine 5'-diphosphates: potent mechanism-based inhibitors of ribonucleotide reductase. *J. Med. Chem.*, 34, 1879-1884.
[198] Khayat AS, Antunes LM, Guimaraes AC, Bahia MO, Lemos JA, Cabral IR, Lima PD, Amorim MI, Cardoso PC, Smith MA, Santos RA, Burbano RR. 2006. Cytotoxic and genotoxic monitoring of sickle cell anaemia patients treated with hydroxyurea. *Clin. Exp. Med.*, 6, 33-37.
[199] Torrents E, Sahlin M, Biglino D, Gräslund A, Sjöberg B-M. 2005. Efficient growth inhibition of *Bacillus anthracis* by knocking out the ribonucleotide reductase tyrosyl radical. *Proc. Natl. Acad. Sci. USA*, 102, 17946-17951.
[200] Elford HL, Wampler GL, van't RB. 1979. New ribonucleotide reductase inhibitors with antineoplastic activity. *Cancer Res.*, 39, 844-851.
[201] Mayhew CN, Sumpter R, Inayat M, Cibull M, Phillips JD, Elford HL, Gallicchio VS. 2005. Combination of inhibitors of lymphocyte activation

(hydroxyurea, trimidox, and didox) and reverse transcriptase (didanosine) suppresses development of murine retrovirus-induced lymphoproliferative disease. *Antiviral. Res.*, 65, 13-22.

[202] Jang M, Cai L, Udeani GO, Slowing KV, Thomas CF, Beecher CW, Fong HH, Farnsworth NR, Kinghorn AD, Mehta RG, Moon RC, Pezzuto JM. 1997. Cancer chemopreventive activity of resveratrol, a natural product derived from grapes. *Science,* 275, 218-220.

[203] Fontecave M, Lepoivre M, Elleingand E, Gerez C, Guittet O. 1998. Resveratrol, a remarkable inhibitor of ribonucleotide reductase. *FEBS Lett.*, 421, 277-279.

[204] Matsuoka A, Lundin C, Johansson F, Sahlin M, Fukuhara K, Sjöberg B-M, Jenssen D, Önfelt A. 2004. Correlation of sister chromatid exchange formation through homologous recombination with ribonucleotide reductase inhibition. *Mutat. Res.*, 547, 101-107.

[205] Pötsch S, Drechsler H, Liermann B, Gräslund A, Lassmann G. 1994. p-Alkoxyphenols, a new class of inhibitors of mammalian R2 ribonucleotide reductase: possible candidates for antimelanotic drugs. *Mol. Pharmacol.*, 45, 792-796.

[206] Roy B, Lepoivre M, Henry Y, Fontecave M. 1995. Inhibition of ribonucleotide reductase by nitric oxide derived from thionitrites: reversible modifications of both subunits. *Biochemistry*, 34, 5411-5418.

[207] Roy B, Guittet O, Beuneu C, Lemaire G, Lepoivre M. 2004. Depletion of deoxyribonucleoside triphosphate pools in tumor cells by nitric oxide. *Free Radic. Biol. Med.*, 36, 507-516.

[208] Nyholm S, Mann GJ, Johansson AG, Bergeron RJ, Gräslund A, Thelander L. 1993. Role of ribonucleotide reductase in inhibition of mammalian cell growth by potent iron chelators. *J. Biol. Chem.*, 268, 26200-26205.

[209] Cooper CE, Lynagh GR, Hoyes KP, Hider RC, Cammack R, Porter JB. 1996. The relationship of intracellular iron chelation to the inhibition and regeneration of human ribonucleotide reductase. *J. Biol. Chem.*, 271, 20291-20299.

[210] Breidbach T, Krauth-Siegel RL, Steverding D. 2000. Ribonucleotide reductase is regulated via the R2 subunit during the life cycle of Trypanosoma brucei. *FEBS Lett*, 473, 212-216.

[211] Breidbach T, Scory S, Krauth-Siegel RL, Steverding D. 2002. Growth inhibition of bloodstream forms of Trypanosoma brucei by the iron chelator deferoxamine. *Int. J. Parasitol.*, 32, 473-479.

[212] Gobin J, Moore CH, Reeve JR, Jr., Wong DK, Gibson BW, Horwitz MA. 1995. Iron acquisition by Mycobacterium tuberculosis: isolation and

characterization of a family of iron-binding exochelins. *Proc. Natl. Acad. Sci. USA*, 92, 5189-5193.
[213] Hodges YK, Antholine WE, Horwitz LD. 2004. Effect on ribonucleotide reductase of novel lipophilic iron chelators: the desferri-exochelins. *Biochem. Biophys. Res. Commun.*, 315, 595-598.
[214] Shao J, Zhou B, Zhu L, Qiu W, Yuan YC, Xi B, Yen Y. 2004. In vitro characterization of enzymatic properties and inhibition of the p53R2 subunit of human ribonucleotide reductase. *Cancer Res.*, 64, 1-6.
[215] Dutia BM, Frame MC, Subak-Sharpe JH, Clark WN, Marsden HS. 1986. Specific inhibition of herpesvirus ribonucleotide reductase by synthetic peptides. *Nature,* 321, 439-441.
[216] Cohen EA, Gaudreau P, Brazeau P, Langelier Y. 1986. Specific inhibition of herpesvirus ribonucleotide reductase by a nonapeptide derived from the carboxy terminus of subunit 2. *Nature*, 321, 441-443.
[217] Duan JM, Liuzzi M, Paris W, Lambert M, Lawetz C, Moss N, Jaramillo J, Gauthier J, Deziel R, Cordingley MG. 1998. Antiviral activity of a selective ribonucleotide reductase inhibitor against acyclovir-resistant herpes simplex virus type 1 in vivo. *Antimicrob. Agents Chemother*, 42, 1629-1635.
[218] Matkovic-Calogovic D, Loregian A, D'Acunto MR, Battistutta R, Tossi A, Palu G, Zanotti G. 1999. Crystal structure of the B subunit of Escherichia coli heat-labile enterotoxin carrying peptides with anti-herpes simplex virus type 1 activity. *J. Biol. Chem.*, 274, 8764-8769.
[219] Hofer A, Schmidt PP, Gräslund A, Thelander L. 1997. Cloning and characterization of the R1 and R2 subunits of ribonucleotide reductase from *Trypanosoma brucei. Proc. Natl. Acad. Sci. USA*, 94, 6959-6964.
[220] Aurelian L, Smith CC. 2000. Herpes simplex virus type 2 growth and latency reactivation by cocultivation are inhibited with antisense oligonucleotides complementary to the translation initiation site of the large subunit of ribonucleotide reductase (RR1). *Antisense Nucleic. Acid. Drug Dev.*, 10, 111-116.
[221] Lee Y, Vassilakos A, Feng N, Lam V, Xie H, Wang M, Jin H, Xiong K, Liu C, Wright J, Young A. 2003. GTI-2040, an antisense agent targeting the small subunit component (R2) of human ribonucleotide reductase, shows potent antitumor activity against a variety of tumors. *Cancer Res.,* 63, 2802-2811.
[222] Heidel JD, Yu Z, Liu JY, Rele SM, Liang Y, Zeidan RK, Kornbrust DJ, Davis ME. 2007. Administration in non-human primates of escalating intravenous doses of targeted nanoparticles containing ribonucleotide reductase subunit M2 siRNA. *Proc. Natl. Acad. Sci. USA*.

[223] Cao MY, Lee Y, Feng NP, Xiong K, Jin H, Wang M, Vassilakos A, Viau S, Wright JA, Young AH. 2003. Adenovirus-mediated ribonucleotide reductase R1 gene therapy of human colon adenocarcinoma. *Clin. Cancer Res.*, 9, 4553-4561.

[224] Yen Y. 2003. Ribonucleotide reductase subunit one as gene therapy target: commentary re: M-Y. Cao et al., Adenovirus-mediated ribonucleotide reductase R1 gene therapy of human colon adenocarcinoma. *Clin. Cancer Res.*, 9: 4304-4308, 2003. *Clin. Cancer Res.*, 9, 4304-4308.

[225] Smith SL, Douglas KT. 1989. Stereoselective, strong inhibition of ribonucleotide reductase from E. coli by cisplatin. *Biochem. Biophys. Res. Commun.*, 162, 715-723.

[226] Chiu CS, Chan AK, Wright JA. 1992. Inhibition of mammalian ribonucleotide reductase by cis-diamminedichloroplatinum(II). *Biochem. Cell Biol.*, 70, 1332-1338.

[227] Giovannetti E, Mey V, Danesi R, Mosca I, Del Tacca M. 2004. Synergistic cytotoxicity and pharmacogenetics of gemcitabine and pemetrexed combination in pancreatic cancer cell lines. *Clin. Cancer Res.*, 10, 2936-2943.

[228] Kjøller Larsen I, Cornett C, Karlsson M, Sahlin M, Sjöberg B-M. 1992. Caracemide, a site-specific irreversible inhibitor of protein R1 of Escherichia coli ribonucleotide reductase. *J. Biol. Chem.*, 267, 12627-12631.

[229] Manfredini S, Baraldi PG, Durini E, Vertuani S, Balzarini J, DeClercq E, Karlsson A, Buzzoni V, Thelander L. 1999. 5'-phosphoramidates and 5'-diphosphates of 2'-O-allyl-beta-D-arabinofuranosyluracil, -cytosine, and -adenine: Inhibition of ribonucleotide reductase. *J. Med. Chem.*, 42, 3243-3250.

[230] Yee KW, Cortes J, Ferrajoli A, Garcia-Manero G, Verstovsek S, Wierda W, Thomas D, Faderl S, King I, O'Brien S M, Jeha S, Andreeff M, Cahill A, Sznol M, Giles FJ. 2006. Triapine and cytarabine is an active combination in patients with acute leukemia or myelodysplastic syndrome. *Leuk. Res.*, 30, 813-822.

[231] Gandhi V, Keating MJ, Bate G, Kirkpatrick P. 2006. Nelarabine. *Nat. Rev. Drug Discov.*, 5, 17-18.

[232] Huang P, Chubb S, Plunkett W. 1990. Termination of DNA synthesis by 9-beta-D-arabinofuranosyl-2-fluoroadenine. A mechanism for cytotoxicity. *J. Biol. Chem.*, 265, 16617-16625.

[233] Parker WB, Shaddix SC, Chang CH, White EL, Rose LM, Brockman RW, Shortnacy AT, Montgomery JA, Secrist Jd, Bennett LJ. 1991. Effects of 2-

chloro-9-(2-deoxy-2-fluoro-beta-D-arabinofuranosyl)adenine on K562 cellular metabolism and the inhibition of human ribonucleotide reductase and DNA polymerases by its 5'-triphosphate. *Cancer Res.,* 51, 2386-2394.

[234] Faderl S, Gandhi V, Keating MJ, Jeha S, Plunkett W, Kantarjian HM. 2005. The role of clofarabine in hematologic and solid malignancies--development of a next-generation nucleoside analog. *Cancer,* 103, 1985-1995.

[235] Atamna H, Paler-Martinez A, Ames BN. 2000. N-t-butyl hydroxylamine, a hydrolysis product of alpha-phenyl-N-t-butyl nitrone, is more potent in delaying senescence in human lung fibroblasts. *J. Biol. Chem.,* 275, 6741-6748.

[236] Larsson Birgander P, Kasrayan A, Sjöberg B-M. 2004. Mutant R1 proteins from Escherichia coli class Ia ribonucleotide reductase with altered responses to dATP inhibition. *J. Biol. Chem.,* 279, 14496-14501.

[237] Persson AL, Eriksson M, Katterle B, Pötsch S, Sahlin M, Sjöberg B-M. 1997. A new mechanism-based radical intermediate in a mutant R1 protein affecting the catalytically essential Glu(441) in *Escherichia coli* ribonucleotide reductase. *J. Biol. Chem.,* 272, 31533-31541.

[238] Kasrayan A, Persson AL, Sahlin M, Sjöberg B-M. 2002. The conserved active site asparagine in class I ribonucleotide reductase is essential for catalysis. *J. Biol. Chem.,* 277, 5749-5755.

[239] Ekberg M, Sahlin M, Eriksson M, Sjöberg B-M. 1996. Two conserved tyrosine residues in protein R1 participate in an intermolecular electron transfer in ribonucleotide reductase. *J. Biol. Chem.,* 271, 20655-20659.

[240] Larsson Birgander P, Bug S, Kasrayan A, Dahlroth SL, Westman M, Gordon E, Sjöberg B-M. 2005. Nucleotide-dependent formation of catalytically competent dimers from engineered monomeric ribonucleotide reductase protein R1. *J. Biol. Chem.,* 280, 14997-15003.

[241] Lycksell PO, Sahlin M. 1995. Demonstration of segmental mobility in the functionally essential carboxyl terminal part of ribonucleotide reductase protein R2 from *Escherichia coli. FEBS Lett.,* 368, 441-444.

[242] Andersson J, Bodevin S, Westman M, Sahlin M, Sjöberg B-M. 2001. Two active site asparagines are essential for the reaction mechanism of the class III anaerobic ribonucleotide reductase from bacteriophage T4. *J. Biol. Chem.,* 276, 40457-40463.

[243] Bodevin S, Westman M, Wincent E, el-Koreiby N, Stålberg T, Sahlin M, Logan DT, Sjöberg B-M. 2007. The active site residues Phe194, Met288, and Arg291 are essential for catalysis by class III anaerobic ribonucleotide

reductase from bacteriophage T4. Manuscript in preparation, (see Chapter 10).

[244] Reichard P, Eliasson R, Ingemarson R, Thelander L. 2000. Cross-talk between the allosteric effector-binding sites in mouse ribonucleotide reductase. *J. Biol. Chem.*, 275, 33021-33026.

[245] Chabes AL, Pfleger CM, Kirschner MW, Thelander L. 2003. Mouse ribonucleotide reductase R2 protein: a new target for anaphase-promoting complex-Cdh1-mediated proteolysis. *Proc. Natl. Acad. Sci. USA*, 100, 3925-3929.

[246] Domkin V, Thelander L, Chabes A. 2002. Yeast DNA damage-inducible Rnr3 has a very low catalytic activity strongly stimulated after the formation of a cross-talking Rnr1/Rnr3 complex. *J. Biol. Chem.*, 277, 18574-18578.

[247] Griffig J, Koob R, Blakley RL. 1989. Mechanisms of inhibition of DNA synthesis by 2-chlorodeoxyadenosine in human lymphoblastic cells. *Cancer Res.*, 49, 6923-6928.

[248] Kjøller Larsen I, Sjöberg B-M, Thelander L. 1982. Characterization of the active site of ribonucleotide reductase of *Escherichia coli*, bacteriophage T4 and mammalian cells by inhibition studies with hydroxyurea analogues. *Eur. J. Biochem.*, 125, 75-81.

[249] Morgan BD, O'Neill T, Dewey DL, Galpine AR, Riley PA. 1981. Treatment of malignant melanoma by intravascular 4-hydroxyanisole. *Clin. Oncol.*, 7, 227-234.

[250] Finch RA, Liu MC, Cory AH, Cory JG, Sartorelli AC. 1999. Triapine (3-aminopyridine-2-carboxaldehyde thiosemicarbazone; 3-AP): an inhibitor of ribonucleotide reductase with antineoplastic activity. *Adv. Enzyme Regul.*, 39, 3-12.

[251] Iwasaki H, Huang P, Keating MJ, Plunkett W. 1997. Differential incorporation of ara-C, gemcitabine, and fludarabine into replicating and repairing DNA in proliferating human leukemia cells. *Blood*, 90, 270-278.

Index

#

3D, 12

A

access, v, 8
acid, 14, 16, 34, 43, 44, 53, 65, 73, 80
Actinobacteria, 30
activase, 6, 8, 21, 25
activation, 28, 37, 79, 82
active site, 3, 4, 7, 14, 19, 21, 40, 41, 44, 49, 50, 52, 54, 58, 79, 86, 87
acute, 56, 59, 85
acute leukemia, 56, 59, 85
acute myeloid leukemia, 59
adaptation, 36
adenine, 43, 59, 85, 86
adenocarcinoma, 52, 57, 82, 85
adenoma, 78
adenovirus, 57
administration, 56
ADP, 5, 26
aerobic, 25, 28, 30, 31, 69, 75
agent, 56, 57, 84
agents, 12, 69
AIDS, 53
air, 20, 21, 40, 41
alkylation, 52

allele, 77
allosteric, 4, 5, 7, 8, 10, 11, 12, 13, 21, 23, 32, 34, 39, 41, 59, 67, 68, 71, 72, 87
alpha, 86
alternative, 55, 69
ambiguity, 59
amino, 11, 14, 16, 34, 44, 65, 80
amino acid, 11, 14, 16, 34, 44, 65, 80
amino acid side chains, 44
amino acids, 11
anaemia, 82
anaerobic, 6, 7, 8, 19, 20, 25, 27, 28, 30, 31, 66, 68, 69, 70, 72, 74, 75, 78, 86
anaerobic bacteria, 72
analog, 86
anemia, 53
animal models, 56
animals, 1
anthrax, 61
anticancer, 81
anticancer drug, 81
antineoplastic, 82, 87
antisense, 48, 57, 84
antisense oligonucleotides, 57, 84
antitumor, 82, 84
Archaea, 1, 7, 8, 9, 23, 25, 34
artificial, 12
asparagines, 86
assimilation, 74

ATP, 5, 11, 13, 21, 22, 26, 32, 33, 34, 35, 65, 67, 76
attacks, 47, 53
atypical, 13, 72
availability, 27, 28, 33, 37

B

Bacillus, 25, 30, 32, 53, 75, 82
Bacillus subtilis, 30, 75
Bacillus thuringiensis, 25
bacteria, 10, 11, 12, 23, 24, 25, 28, 29, 30, 31, 34, 35, 39, 53, 72, 73, 76
bacterial, 1, 15, 24, 30, 32, 34, 72, 76
bacteriophage, 15, 44, 68, 70, 75, 86, 87
bacteriophages, v, 9, 10
bacterium, 28, 29, 30, 31, 35, 61, 71
behavior, 19
beta, 85, 86
binding, 1, 5, 12, 13, 14, 15, 16, 20, 22, 26, 28, 33, 34, 35, 36, 41, 49, 50, 55, 56, 67, 69, 70, 75, 79, 81, 84, 87
biochemical, 15, 49, 56, 61, 79
bioinformatics, 27, 61
biophysical, 49
biosynthesis, 34, 74
biosynthetic pathways, 69
birds, 10
black, 7
blocks, 2, 66
blood, 55, 58
bloodstream, 83
bonds, 50
bone, 52, 53
bone marrow, 52, 53
Boston, 74
bottleneck, 47
budding, 35
building blocks, 1, 66

C

C. diphtheriae, 31
C. thermocellum, 15

C. trachomatis, 13, 15
cAMP, 27
cancer, 39, 47, 48, 51, 52, 54, 57, 58, 59, 63, 66, 77, 78, 81, 82, 83, 84, 85, 86, 87
cancer treatment, 51
cancers, 52
candidates, 47, 83
capacity, 20, 35
carboxyl, 79, 86
carcinoma, 52, 77, 78
catalysis, 5, 20, 68, 71, 86
catalytic, v, 11, 12, 13, 15, 19, 21, 22, 70, 79, 87
catalytic activity, 79, 87
cattle, 61
cDNA, 65, 77
cell, 21, 24, 25, 26, 28, 31, 36, 39, 43, 45, 47, 48, 51, 52, 53, 55, 56, 57, 58, 59, 69, 73, 78, 81, 82, 83, 85
cell culture, 45
cell cycle, 24, 25, 26, 28, 31, 36, 55, 69, 73
cell growth, 83
cell line, 47, 48, 52, 58, 78, 81, 85
cell lines, 47, 48, 52, 58, 78, 81, 85
chelators, 47, 54, 55, 83, 84
chemical, 25, 53, 61
chemicals, 27, 47
chemopreventive, 83
chemotherapy, 54, 66
Chlamydia, 11, 15, 40, 70
Chlamydia trachomatis, 11, 40, 70
chlorine, 49
chromatid, 83
chromosome, 10, 14, 24
cis, 26, 85
cisplatin, 43, 57, 58, 85
cladribine, 59
classes, v, 3, 4, 5, 11, 12, 19, 22, 24, 29, 30, 32, 33, 34, 35, 67, 72, 75
classical, 9, 10
classification, 2, 3
classified, 8, 13
cleavage, 3, 7, 8
clinical, v, 2, 45, 47, 51, 56, 58, 59, 61
clinical trial, v, 2, 45, 47, 51

Index

clinical trials, v, 2, 45, 47, 51
clones, 65
cloning, 1, 25, 74, 77
clusters, 35
cobalamin, 3, 7
codes, 33
coding, 10, 25, 32, 39, 57, 75
codon, 28
coenzyme, 3
colon, 57, 85
colorectal, 78
combat, 47
combination therapy, 53
communication, 17, 27, 31, 51
competition, 56
competitor, 52
complement, 1, 31
complementary, 84
complications, 47
components, 16, 25, 34, 73, 81
composite, 14
composition, 13
compounds, 27, 47, 53
concentration, 52, 58
conformational, 32
Congress, iii
consensus, 27, 28, 34, 40, 44
conservation, 16
control, 1, 23, 28, 32, 76
controlled, v, 30, 75
conversion, 26
coordination, 34, 79
correlation, 54
Corynebacterium diphtheriae, 24, 71, 72, 75
couples, 73
coupling, 27
covalent, 49
crops, 61
cross-talk, 87
crystal, 20, 56, 67
crystal structure, 20, 56, 67
crystal structures, 20
crystallization, 49
C-terminal, 12, 13, 14, 20, 21
C-terminus, 56

cues, v, 24, 29, 35
culture, 45
curing, 47
cyclodextrin, 57
cysteine, 3, 4, 7, 12, 13, 21, 40, 41, 44, 58, 67, 68
cysteine residues, 4, 67, 68
cytoplasm, 37
cytosine, 49, 85
cytotoxic, 51, 58, 59, 82
cytotoxicities, 59
cytotoxicity, 85

D

D. melanogaster, 52
database, v, 8, 9, 11, 70, 78
death, 48, 73
decay, 48, 49, 54
defects, 39
defense, 48, 54
defense mechanisms, 48, 54
deficiency, 30, 31, 52, 82
degenerate, 1
degradation, 36, 41, 56, 57
degree, 16
delivery, 56, 57
denitrifying, 75
deoxyribonucleic acid, 73
deoxyribonucleotides, v, 1, 15, 21, 65
dependant, 34
diamond, 13, 16
diffraction, 78
diiron cluster, 79
dimer, 4, 5, 10, 40, 67
distribution, 8, 9, 30
disulfide, 21
diversification, 20, 21
diversity, 24
DNA, v, 1, 10, 14, 19, 23, 25, 26, 27, 28, 30, 32, 34, 35, 36, 37, 39, 43, 48, 52, 54, 57, 59, 61, 66, 68, 69, 70, 71, 73, 74, 77, 85, 86, 87
DNA damage, 10, 27, 28, 36, 37, 66, 69, 70, 77, 87

DNA ligase, 59
DNA polymerase, 59, 86
DNA repair, 23, 30, 32, 77
DNA sequencing, 39
down-regulation, 30
Drosophila, 30
drug delivery, 56
drug design, 53, 66
drugs, 2, 45, 47, 48, 51, 53, 54, 55, 57, 58, 59, 61, 83
duplication, 11, 13, 24, 67

EST, 78
Eukaryota, 1, 7, 9
eukaryote, 68
eukaryotes, v, 5, 8, 9, 10, 34, 39
eukaryotic, 1, 5, 8, 10, 44, 49, 61, 81
evidence, 26, 35, 37, 80, 81
evolution, 9, 19, 20, 22, 68, 71
evolutionary, 7, 19, 68, 71
expert, iii
exponential, 29, 31, 32
exposure, 73

E

E. coli, 13, 15, 16, 20, 25, 26, 27, 28, 35, 39, 44, 48, 49, 52, 54, 58, 65, 73, 80, 85
E2F, 69
efficacy, 61
electron, 3, 44, 78, 80, 86
electronic, iii
electrons, 21
electrostatic, iii
elongation, 25, 48
encoding, 16, 24, 32, 35, 39, 65, 66, 70, 72, 73, 75, 77
endogenous, 54
endonuclease, 13, 70
ENDOR, 78
engineering, 44, 68
enterococci, 61
environment, 15, 35, 40, 61
environmental, v, 24, 29, 31, 35
environmental conditions, 31, 35
enzymatic, 12, 44, 77, 79, 84
enzymatic activity, 12, 79
enzyme, v, 1, 5, 10, 11, 12, 14, 15, 17, 19, 20, 21, 23, 27, 28, 29, 30, 31, 33, 35, 39, 40, 41, 42, 44, 47, 48, 54, 55, 66, 68, 69, 71, 73, 80
enzymes, 1, 11, 12, 19, 20, 61, 69, 70, 72, 75
EPR, 15, 78
equilibrium, 36
Escherichia coli, 1, 6, 40, 41, 65, 66, 67, 68, 69, 72, 73, 74, 76, 77, 78, 79, 80, 81, 84, 85, 86, 87

F

family, 8, 11, 20, 21, 24, 25, 72, 75, 84
feedback, 10, 59
feedback inhibition, 59
fibroblasts, 86
fludarabine, 59, 87
fluoride, 49
fluorides, 49
fluorine, 49, 59
folding, 36
free radical, 66, 67, 78, 80
fumarate, 28
fungi, 1, 10
Fur, 27, 28, 31, 74
Fusarium, 61
fusion, 56, 72

G

gastric, 77
GDP, 5
gene, 1, 5, 7, 8, 9, 10, 12, 13, 16, 24, 27, 29, 30, 31, 32, 33, 35, 36, 39, 44, 48, 57, 66, 68, 69, 72, 73, 74, 76, 77, 78, 81, 85
gene amplification, 81
gene expression, 1, 30, 73, 76
gene therapy, 48, 57, 85
gene transfer, 9
generation, 5, 44, 54, 68, 80, 81, 86
genes, v, 1, 5, 7, 8, 9, 10, 12, 13, 14, 16, 23, 24, 25, 26, 27, 28, 29, 30, 31, 32, 33, 34,

Index

35, 36, 37, 39, 61, 67, 69, 70, 72, 73, 74, 75, 76, 81
genetic, 15, 23, 61, 78
genome, 1, 10, 24, 25, 70, 71, 72
genome sequencing, 1, 25
genomes, 1, 8, 9, 10, 11, 16, 34, 68
genomic, v, 1, 26, 28, 72
genomics, 34
genotoxic, 82
germination, 32
Gibberella, 10
glutathione, 27, 73
glycine, 20, 34, 72, 78
glycyl, v, 3, 7, 14, 15, 20, 21, 41, 44, 67, 69, 72, 78
gracilis, 68
Gram-positive, 24, 31
grants, 63
grapes, 53, 83
groups, 2, 9, 11, 25, 49
growth, v, 24, 27, 28, 29, 30, 31, 32, 34, 57, 61, 74, 75, 76, 77, 82, 83, 84
growth inhibition, 82

H

H_1, 6
H_2, 6
half-life, 55
halogenated, 48
halogenation, 59
heat, 56, 84
Helicobacter pylori, 24
hematologic, 86
herpes, 54, 70, 84
herpes simplex, 54, 70, 84
herpes simplex virus type 1, 70, 84
heterodimer, 10, 35, 37, 69
Hilbert, 75
HIV, 53
holoenzyme, 10, 15, 36, 37
homeostasis, 74
homology, 15, 35, 65
homolytic, 3, 7
horizontal gene transfer, 9

host, 53, 56, 57
human, 15, 51, 52, 57, 77, 78, 81, 82, 83, 84, 85, 86, 87
human leukemia cells, 87
human lung fibroblasts, 86
hydrogen, 3, 8, 50, 74, 79, 80
hydrogen abstraction, 3, 8
hydrogen bonds, 50
hydrogen peroxide, 74
hydrolysis, 86
hydrophobic, 44
hydroxyl, 49
hypothesis, 11, 20, 21, 28, 30, 31

I

id, 25
identification, 73, 74, 75
immunodeficiency, 53
in situ, 54
in vitro, 31, 43, 47, 48, 49, 53, 69
in vivo, 47, 48, 56, 58, 73, 74, 84
inactivation, 49, 80, 81
inactive, 28, 36, 55, 57
induction, 27, 28, 31
infection, 28, 30, 56, 65
infections, 47, 54, 55, 56, 61
infectious, 28, 30, 61
inhibition, 25, 27, 33, 37, 40, 41, 47, 56, 59, 61, 73, 82, 83, 84, 85, 86, 87
inhibitor, 2, 5, 42, 43, 48, 49, 50, 51, 52, 54, 59, 77, 81, 83, 84, 85, 87
inhibitors, v, 25, 42, 48, 49, 54, 57, 58, 59, 66, 73, 82, 83
inhibitory, 47, 51, 59, 82
inhibitory effect, 51, 59
initiation, 25, 26, 53, 84
injection, 79
injury, iii
insertion, 13, 16, 77
insight, 1
instability, 58
interaction, 40, 41, 42, 48, 56, 68, 79
interactions, 39
interface, 5, 12

interference, 48
intermolecular, 86
intravascular, 87
intravenous, 84
introns, 8
inversion, 25
inverted repeats, 31
ions, 35
iron, 8, 15, 16, 24, 27, 28, 31, 35, 40, 43, 47, 53, 54, 55, 66, 67, 69, 71, 74, 78, 79, 83, 84
irradiation, 72
isolation, 39, 66, 83

J

Jordan, 67, 68, 71, 72, 73, 74, 75
Jung, 78

K

K-12, 72, 74
kinase, 37, 52, 59, 77, 82
kinases, 48, 51
kinetics, 49, 54
King, 85
knockout, 30, 31, 35, 39

L

Lactobacillus, 12, 41, 81
latency, 57, 84
lead, 23, 61
leukemia, 52, 54, 56, 59, 85, 87
leukemia cells, 54, 87
life cycle, 55, 83
ligand, 17, 24, 32, 33, 78, 79
ligands, 15, 16
limitation, 27, 28, 31, 32, 76
links, 8
lipophilic, 55, 84
literature, 6, 59
localization, 37, 77
location, 8
locus, 16, 39

long distance, 44
lung, 52, 58, 81, 86
lung cancer, 52, 58, 81
lungs, 31
lymphoblast, 54
lymphocyte, 82
lymphomas, 57

M

M1, 6, 51, 65, 77, 78
macrophage, 54
macrophages, 28, 75
magnetic, iii
malignant, 53, 87
malignant melanoma, 53, 87
Mammalian, 69
mammalian cell, 48, 55, 77, 81, 83, 87
mammalian cells, 48, 55, 77, 81, 87
mammals, 1, 10, 57
manganese, 31
mapping, 1, 14
marrow, 52, 53
mechanical, iii
melanoma, 53, 57, 87
metabolic, 72, 75, 76
metabolism, 23, 34, 51, 76, 86
metabolite, 32
metabolites, 51
metastatic, 57
methionine, 74
methylene, 81
mice, 31
microaerophilic, 28, 31
microarray, 27, 74
microorganism, 34
microorganisms, 2, 15, 17, 66
mimicking, 61
mitosis, 41
mitotic, 66
mobility, 86
model system, 30
models, 56
modules, 20
molecules, 49, 54

monomer, 67
monomeric, 86
Moon, 83
morphological, 77
Mössbauer, 66
mouse, 1, 10, 39, 44, 52, 54, 57, 59, 61, 65, 69, 71, 77, 78, 80, 81, 87
mRNA, 12, 57
mutagenesis, 16, 71, 79, 80
mutant, 1, 27, 28, 31, 35, 77, 78, 79, 86
mutant proteins, 79
mutants, 25, 28, 30, 32, 39, 77, 78, 79
mutation, 23, 26, 32, 39, 41, 44, 77
mutations, 8, 25, 39, 40, 44
Mycobacterium, 6, 16, 24, 75, 83
myeloid, 59

N

N-acety, 43
naming, 15
nanoparticles, 84
natural, 49, 83
nematodes, 10
neurotoxic, 58, 59
neurotoxic effect, 59
New York, ii, iii
nitrate, 28
Nitric oxide (NO), 27, 54, 74, 83
nitric oxide synthase, 54
nitrogen, 27, 74, 80
nitrosative stress, 28
Nixon, 65
non-human, 84
non-human primates, 84
non-small cell lung cancer, 81
normal, 13, 27, 29, 32, 36, 50
normal conditions, 36
NRC, 23, 72
N-terminal, 5, 11, 13, 20, 21, 35
nuclear, 37, 78
nucleolus, 37
nucleotide sequence, 1, 8, 70, 72
nucleotides, 28

O

observations, 33
oligomeric, 4
oligomerization, 11
oligonucleotides, 57, 84
operon, 1, 5, 8, 11, 24, 25, 28, 31, 32, 35, 65, 73, 74, 76
oral, 82
organism, v, 8, 9, 23, 25, 30, 35, 61
organization, 7, 23, 72
oxidation, 3, 5, 16, 44
oxidative, 27, 28, 31, 73
oxidative damage, 73
oxidative stress, 27, 31
oxide, 27, 54, 74, 83
oxygen, v, 5, 7, 8, 16, 19, 28, 30, 31, 32, 44, 54, 71, 74, 75, 76, 78, 79

P

p53, 6, 10, 69, 77
pancreatic, 58, 85
pancreatic cancer, 85
paramagnetic, 68
parasite, 47, 54, 55
parasites, 17, 54
parasitic infection, 61
Paris, 84
patents, 47
pathogenic, 53
pathways, 27, 69
patients, 56, 57, 59, 78, 82, 85
PCR, 29
pemetrexed, 85
peptide, 1, 56
peptides, 56, 84
peroxide, 74
personal, 17, 27, 31, 51
personal communication, 17, 27, 31, 51
pests, 61
pH, 80
phage, 11, 12, 13, 65
pharmacogenetics, 85

phenotype, 14, 39, 77
phenylalanine, 15
phosphorylation, 36, 41, 48
phylogenetic, 11, 12, 34, 76
phylogenetic tree, 11, 12
phylogeny, 9
physiological, 5, 12, 58, 76
plants, 1
plasmid, 1
plasticity, 12
polymerase, 48, 59
polymorphism, 39, 78
polymorphisms, 78
polypeptide, 5, 7, 10, 35
polypeptides, 4, 5, 8, 10, 70
pools, 31, 53, 54, 59, 69, 83
preparation, iii, 87
primates, 84
production, v, 1
profit, 49
program, 63
prokaryotes, v, 5, 8, 23
prokaryotic, 9
proliferation, 39, 47
promote, 61
promoter, 25, 26, 27, 28, 31, 34, 35, 69, 72, 76
promoter region, 25, 26, 27, 28, 31, 34, 35, 72, 76
propagation, 80
property, iii
proteases, 56
protection, 74
protein, 4, 5, 8, 10, 11, 12, 13, 14, 15, 16, 19, 20, 21, 24, 25, 26, 28, 31, 33, 34, 35, 37, 39, 44, 48, 49, 56, 57, 58, 66, 67, 68, 69, 70, 71, 72, 73, 76, 77, 78, 79, 80, 81, 85, 86, 87
protein engineering, 68
protein family, 21
protein function, 26
protein sequence, 8, 16
protein-protein interactions, 39
proteins, 5, 6, 11, 12, 13, 14, 15, 16, 20, 21, 22, 35, 40, 44, 48, 65, 68, 69, 70, 71, 72, 77, 79, 86

proteobacteria, 11, 12, 24, 34, 72
Proteobacteria, 9, 14
proteolysis, 87
proteomes, 72
prototype, 13, 48
proximal, 69
Pseudomonas, 6, 9, 30, 41, 67, 69, 75
Pseudomonas aeruginosa, 6, 9, 41, 67, 69
public, 1, 6, 8, 11
pyrimidine, 53
pyruvate, 20

Q

question mark, 29
Quinones, 74

R

radiation, 23
radical, v, 3, 4, 5, 7, 14, 15, 16, 20, 21, 22, 27, 35, 40, 41, 42, 44, 47, 48, 49, 52, 53, 54, 55, 56, 66, 67, 68, 69, 71, 72, 78, 79, 80, 81, 82, 86
radical formation, 36, 81
radical mechanism, 69
radical reactions, 66
range, 80
rat, 81
reaction mechanism, v, 4, 19, 20, 39, 40, 44, 86
reactive nitrogen, 27, 74
reactivity, 53
reading, 12, 31, 34, 72
real-time, 29
recombination, 12, 72, 83
redox, 13, 21, 44, 55, 66, 67, 80
redox-active, 13, 21, 44, 67, 80
reductases, 65, 66, 71, 73, 75, 76, 81
reduction, 3, 5, 23, 26, 28, 54, 67, 68, 69, 71, 75, 82
regeneration, 48, 83

regulation, 1, 5, 10, 11, 19, 23, 25, 27, 28, 29, 30, 31, 32, 33, 34, 35, 36, 37, 39, 41, 47, 67, 68, 70, 71, 72, 73, 74, 76
regulators, 1, 25, 27, 74
relationship, 82, 83
relatives, 20
relevance, 15
repair, v, 2, 23, 30, 32, 61, 77
replication, v, 2, 10, 26, 30, 35, 37, 57, 61, 73
repression, 33
repressor, 75
research, 9
residues, 4, 14, 16, 20, 24, 34, 35, 49, 50, 67, 68, 79, 86
resistance, 47, 48, 51, 52, 59, 81, 82
resolution, 63
respiration, 28
Resveratrol, 43, 53, 83
retrovirus, 53, 83
reverse transcriptase, 83
riboflavin, 34
ribonucleotide reductases, 65, 66, 71, 73, 75, 76, 81
ribose, 48, 49
risk, 77
RNA, 19, 32, 33, 48, 51, 71, 76
RNAs, 76

S

S phase, 37, 55
Saccharomyces cerevisiae, 6, 10, 42, 66, 77
Salmonella, 25, 28, 61, 67, 71, 72, 73, 75
scavenger, 27, 43, 52, 54
search, 34, 55
SEARCH, 8
searches, 8
senescence, 86
sensing, 34
separation, 12
sequencing, 1, 8, 25, 28, 39, 72
series, 44, 54
services, iii
shade, 7
sickle cell, 53, 82

sickle cell anemia, 53
side effects, 43, 52, 55, 61
signaling, 54
similarity, 28, 57
single nucleotide polymorphism, 39, 78
siRNA, 57, 84
sister chromatid exchange, 83
sites, 1, 4, 75, 87
sleeping sickness, 55, 56
SNP, 39
sodium, 27
solid tumors, 57
solutions, 58
species, 16, 24, 27, 30, 53, 54, 56, 72, 74, 75
specificity, 4, 7, 8, 12, 13, 19, 21, 68
spectroscopy, 66
S-phase, 36, 37
spore, 32
stability, 5, 40, 79
stages, 28, 32
Staphylococcus, 30, 75
Staphylococcus aureus, 30, 75
stars, 16
starvation, 25
stereospecificity, 54
stoichiometry, 81
strain, 61, 68
strength, 56
streptococci, 61
Streptomyces, 30, 31, 34, 76
stress, 27, 28, 31, 37, 53
subgroups, 14
substitution, 16
substrates, 4, 48
sugar, 59
sulfur, 8, 69
Sun, 69, 72, 78
suppression, 57
suppressor, 10
survival, 30, 66
syndrome, 85
synergistic, 58
synergistic effect, 58
synthesis, 2, 10, 23, 24, 25, 26, 27, 39, 52, 54, 59, 65, 66, 73, 77, 85, 87

T

synthetic, 57, 84
systems, 14, 26, 28, 32, 47, 54, 57, 72

tandem repeats, 77
targets, 51
taxonomy, 8
temperature, 25, 27, 39, 77
tension, 30
theoretical, 49, 66
theory, 8
therapeutics, 57
therapy, v, 2, 48, 51, 52, 53, 55, 56, 57, 58, 59, 81, 85
thioredoxin, 21, 26, 40, 41, 68, 73
three-dimensional, 4, 8, 20, 44
thymidine, 25
time, 3, 21, 25, 27, 29, 31, 79
tissue, 55
topology, 11, 25
toxic, 43, 47, 55, 61
toxic effect, 47
toxic side effect, 43, 55
toxicities, 56
toxicity, 52, 53, 54, 55
toxin, 75
trade, 59
trans, 54
transcript, 32, 74
transcriptase, 83
transcription, 1, 23, 25, 26, 27, 28, 29, 30, 31, 32, 34, 35, 69, 73, 75, 76
transcription factor, 1, 75
transcription factors, 1, 75
transcriptional, 23, 24, 25, 26, 27, 29, 30, 31, 32, 33, 34, 35, 37, 69, 72, 73, 74, 76
transcripts, 32
transfection, 52
transfer, 3, 10, 40, 42, 44, 79, 80, 86
transformation, 57
transition, 19, 28
transitional cell carcinoma, 78
translation, 84
transport, 48, 76
transposon, 16, 71
trial, 82
tryptophan, 78
tuberculosis, 6, 24, 30, 31, 55, 61, 75, 83
tumor, 10, 53, 54, 57, 61, 83
tumor cells, 54, 57, 61, 83
tumors, 57, 84
turnover, 20
tyrosine, 15, 20, 40, 67, 78, 86
tyrosyl radical, v, 3, 4, 5, 15, 16, 20, 35, 40, 42, 44, 47, 48, 49, 52, 54, 55, 56, 66, 71, 78, 79, 82

U

ubiquitin, 41
unclassified, 13, 24
urinary, 78
urinary tract, 78
user-defined, 8
UV, 23, 72, 73
UV exposure, 73
UV irradiation, 72
UV-radiation, 23

V

variable, 24
vector, 57
viral, 10
virulence, 28
virus, 6, 10, 47, 54, 57, 70, 84
virus infection, 47
viruses, v, 1, 9, 10, 44, 56
visualization, 69
vitamin B1, v, 3, 12, 20, 30, 32, 33, 34, 76
vitamin B12, v, 3, 12, 20, 30, 32, 33, 34, 76

W

water, 3, 49, 50
workers, 25, 32

X

x-ray, 78, 79
X-ray crystallography, 79
x-ray diffraction, 78

Y

yeast, v, 1, 10, 23, 35, 66, 69

Z

zinc (Zn), 34, 35, 41, 76